SAY IT LOUDER!

Black Voters, White Narratives,
and Saving Our Democracy

Tiffany D. Cross

AMISTAD

An Imprint of HarperCollinsPublishers

HarperCollins books may be purchased for educational, business,
or sales promotional use. For information, please email the Special
Markets Department at SPsales@harpercollins.com.

FIRST EDITION

Designed by Terry McGrath

Library of Congress Cataloging-in-Publication Data is available
upon request.

ISBN 978–0–06–297676–5
ISBN 978–0–06–297677–2 (Library Edition)

20 21 22 23 24   LSC   10 9 8 7 6 5 4 3 2

For the millions of Black people, both living and dead, whose lived experiences helped to shape and save this democracy since its inception. You are not invisible.

Contents

Foreword: Bearing the Cross

By Michael Eric Dyson

On February 26th, 2020, the nation's highest-ranking Black Congressman dramatically changed the 2020 presidential race when he endorsed the beleaguered former vice president Joe Biden.

"I know Joe. We know Joe. But most importantly, Joe knows us."

Those words radiated the sentiment of older Black Southerners who felt a sense of racial intimacy and political kinship with Biden over the years. After all, their options for acceptable political leadership, save for President Barack Obama, had been few. Clyburn's affirmation was bone deep Black affirmation, too, at least among many Black folks in the region and beyond.

Most of the world was stunned at the outcome of Clyburn's well-placed and signifying words. Well, that is, that part of the world that wasn't Black, that part of the world that failed, or refused, to take Black folk seriously, that part that didn't listen to our thoughts and insights and philosophies from barbershops to beauty salons, from church sanctuaries to mosque meetings, from school halls to street corners, from basketball courts to backyard cookouts. Black folk relentlessly politic, talk politics, politicize their existence, engage issues politically. And yet much of this is missed, overlooked, ignored, discounted, or downright dissed, as if it doesn't matter to the country, as if our interests don't count, as if our views don't affect democracy, or shape the nation's economic or political flow, as if being Black was tangential rather than central to America.

If political observers have failed to do their due diligence, the fourth estate, too, has often been hamstrung by a biased and distorted view of Black life. How many nights have we gazed on white faces and heard white voices interpret Black experience, weigh in on Black voting behavior, or most likely, offer poorly sourced or ill-informed reflections on the complicated political expressions of Black folk where the data is congested and rarely disaggregated? There is no such thing as "the Black vote," any more than there is anything like "the Black community," or "the Black point of view." There are Black voters, there are Black communities, there are Black points of view. Blackness, essentially, historically, and principally, is plural, not singular, is complex, not simple, is better seen in broader rather than narrow scope.

For instance, there are Black folk who endorse gay marriage, and many more, admittedly, who stand against it. There are many progressive Black folks on the political spectrum and many more in the centrist wing too. There are millions of Black folks in urban America, and many, too, who occupy the rural hinterland and rushing backwaters. There are older Black folk in the south who embrace more conservative moments of liberal thought, while some adhere strictly to leftwing beliefs. And there are younger Black folk by the thousands who embrace the democratic socialism of Bernie Sanders even as they spar with their Black peers or elders in every region of the country that favor more moderate views. Black folk also feature measured political figures like Bakari Sellers, the millennial son of Black activist royalty from South Carolina, and Maxine Waters, of the "silent generation," an eighty-one-year-old progressive fire-brand who was awake decades before woke folk got awakened by a political alarm clock a few days ago.

At the same time that the mainstream political and media establishment overlook the heteroglossia of Black languages of social activism and political resistance, it has focused nearly sin-gular, at times excessive, attention on white folk and their fears, feats, and phobias—their desires, desperations, and disappoint-ments, too—as if they exclusively define our democracy. When important white ideas and issues are engaged in exclusion from Black ideas and issues, it makes the white ones seem essential, natural, normal, and definitive, and the Black ones appear ines-sential, abnormal, aberrant, and incidental. When we talk about putative white economic anxiety in a fashion that ignores the persistent economic vulnerability of Black folks and communi-ties, it makes it appear as if Black folk don't suffer the same, if not considerably more, anxieties. It makes it appear that Black economic woes don't shape the nation in the same way, or that they don't impact the conduct or conscience of the country in the same manner.

Or, more bluntly, it makes it seem that what happens to Black folk pales politically and economically in comparison to what happens to white folk. When political actors and media pundits spotlight the suffering of the white poor, underscore the plight of the white middle classes, or focus on the doings of the white wealthy, while ignoring those of their fellow black citizens, it suggests, at best, a harsh dismissal of vast regions of Black thought and experience, and, at worst, a two-tiered, Jim-Crow system of stunted political analysis and stymied social conscience.

Into that unforgiving gap steps the remarkable Tiffany Cross. In all my fifty years of racial resistance, my forty years of political activism, my thirty-five years of scholarship, and my more than thirty years in media commentary, I have rarely met anyone as talented as Tiffany Cross. She tells hard Black truths to often clueless white audiences with panache, humor, insight, sophisticated sass, and enormous erudition. She is a woman warrior with an old soul who's capable of deep empathy for the Black generations that came before her and for those coming up behind. She is deeply learned about the political behavior of Black folk who endured naked American terrorism in the south under Jim Crow and those who endure police terror in the north under the post-Obama Trump racial apocalypse. Cross can specify the complicated character of Black political behavior. She can divine the trends and forces that shape Black belief about the country Black folk have monumentally fashioned. Cross pays serious attention to Black unemployment and ideological variety. She knows well the paradox of Black folk being the cutting edge of political progressivism even as they share the political middle with millions of white folk and others in a way rarely acknowledged.

Tiffany Cross is especially adept at parsing the meanings and modes of white supremacy as they metastasize across the body politic and show up in both our cultural gutters and our political thrones. She insists on television, as she does brilliantly in this magnificent book, that white folk see Black folk and hear Black folk and study the political terrain that Black folk so colorfully and creatively inhabit. She holds the white media to account, takes its pundits to task, for speaking of the working class without stipulating that it's the white working class, the same white working class that went for Donald Trump hook, line, and sinker. As she makes clear in *SAY IT LOUDER!*, so did middle-class and rich white folk, too. But the Black working class hardly fell for the okey doke. That distinction is critical.

In this brilliant, engaging book, Cross disaggregates the data, crunches the numbers, breaks down the factors that constitute Black voting patterns, and identifies Black cultural forces that shape Black communities and the broader American society. Cross also makes clear that poor, middle-class, and rich white folk voted for Trump. The mainstream media and political establishment too often overlook the force of whiteness in dominant culture—the felicities, enticements, seductions, entitlements, promises, and rewards of whiteness. They also ignore how whiteness overwhelms and to a degree counteracts, and yes, at times, transcends class and economic differences. *SAY IT LOUDER!* makes the argument for why this truth must be more rigorously investigated and thoroughly examined. It makes the case that whiteness is not normative, that it cannot exhaust the experiences of Black folk and many others, that it must not set the template for their interpretation or deconstruction.

Tiffany Cross—on television, on social media, and in this important book—crashes the gatekeepers who whitewash the American experience and, as she forcefully argues, destroy American democracy in the process. Black folk and other people of color will soon be the majority in a country where they are discounted, undercounted, counted out, and certainly not counted on to provide political prudence, wise counsel, sure guidance, and a reliable barometer of the nation's political affairs. If the conversations Black folk and other people of color are having are not being heard by the majority, it doesn't mean that those conversations won't have an effect on the nation's course of events, as the Clyburn endorsement clearly showed. It is not only just to include those voices in our other mainstream institutions. It makes good sense, too. Getting, polling, analyzing, interpreting, weighing, and integrating the views of Black folk and others is like gathering a moral census: the distribution of perspectives in an economy of attention says a lot about who and what we are as a nation. Black blood was spilled to shape this nation; Black labor was shared to build its institutions; Black intelligence has remade American democracy. And yet the Black presence in politics and media has been both slighted and erased.

Tiffany Cross brilliantly wears two hats—and pivots between two realities—as both political analyst and media commentator: She seeks to reshape the narratives that condition our awareness of Black voters, and she seeks to offer a vibrant historical context that helps illuminate strategies that increase Black voter participation. She does both extremely well.

Tiffany has consistently borne the Cross to speak the truth

about how Black sacrifice and suffering have transformed American politics. She bears the Cross of cutting against the grain, articulating uncomfortable realities that few others are willing to acknowledge, and boldly expressing Black common sense about matters of state. Tiffany has borne the Cross of contesting the illusory homogeneity of Black life and insisted on the robust diversity and surprising paradoxes of Black worldviews. Tiffany has borne the Cross of insisting that only the plural will do where the singular has been lazily inserted. On the Crossroads of American political destiny, Tiffany has Crossed swords, and been willing to be at Cross purposes, with those who refused to Cross the political street to the other, blacker side.

Consider *SAY IT LOUDER!* a primer in Black political thought and behavior, a Cross training in multiple modes of intellectual and political exercise to enhance our democratic fitness. At this moment, Crossing over to her way of seeing the world might be the greatest Tiffany gift we receive.

Introduction

I f there was a way to yell *at* the television while being *on* television, I would have done it by now. Like you, I have grown increasingly frustrated with the faulty ways some have tried—and failed—to chronicle our current state of affairs. After years of watching much of the media landscape miss the mark when it comes to Black voters, I've got to counter some of these inaccurate political narratives.

Like "Black women were torn between their race and their gender when it came to Barack Obama and Hillary Clinton" or "Black people may not be supporting Mayor Pete Buttigieg because he's gay" or "Black people just decided to stay home in the 2016 election after Obama was no longer on the ticket." (Insert eye roll here.) The perpetual narratives relating to Black people and our politics are formed on, grounded in, and perpetuated by the notion that society functions and spins on the axis of white people. That thinking has led us to this place: on the precipice of a crumbling MAGA mountain in a desperate and frantic state of chaos. (Note all of Black America staring across

1

the chasm with pursed lips and raised eyebrows, screaming, "*We tried to tell you.*") So for the same master narrators, as Toni Morrison referred to them, to lay this democracy's failure at our feet is as preposterous as assigning themselves as its savior.

Mainstream media outlets have long influenced how American civilization, or lack thereof, has functioned. For Black folks to disrupt that narrative, throughout time, we would have to spill blood and lose countless lives. Speaking these truths can make the white power structure uncomfortable. But that's the only way to wrest the narrative away from the master narrator. Ergo, for the sake of all that Black people have survived and endured, I'm comfortable making them *uncomfortable*. Entirely too much about our story is controlled by people willfully ignorant of our American experience. We are storytellers; it is a rich tradition of our lineage. Having someone try and fail to capture our memories is a slow death. And when it comes to our politics, white media gatekeepers are bleaching the Black American experience and killing democracy at the same damn time.

Trying to keep up with the anarchy that the Trump administration brought is like trying to catch confetti. It would be challenging for anyone in the media space to keep up, but this pit of chaos was all by design. It's just like the character Littlefinger said in *Game of Thrones*: "Chaos isn't a pit, it's a ladder." The Trump administration wanted to destroy the pillars of democracy and climb up its ashes to rule the nation. A brief look at the country's racist history could have predicted that we would land here eventually. Given that African Americans have been enslaved in this country for longer than we have been a freed people, we should typically be the first point of reference on all things white supremacy. When the mainstream media refused to call it by name, we shouted it and cursed it, because we live it.

The racism on which America was founded didn't push us away from participating in democracy. It never let us in at the onset. The United States was never founded with Black people in mind. It was as if the framers were designing a house. And even though the bricklayers, plumbers, electricians, and architects—who were forced to endure backbreaking labor without compensation—were also going to be living in this house, their input was never even a thought. These designers could only see their vision, not the people actually bringing it to life. So later, those representing the power structure chose to enforce something they never even considered in the first place: keeping African Americans, the builders of this nation, from participating in representative government. Still, Black people have saved this nation many times fighting with and for the very system designed for our demise. We are not just participants in the American experiment. We are also every bit as much its architects as the framers. Our blood stripes the American flag red, staining the white backdrop. Yet that white backdrop still dominates so much.

Now, bear with me on this journey. I'm a writer and I'm sensitive about my sh*t. I tend to put things in consumable terms, not like an academic who fills pages with abstruse disquisition (not that the two must be mutually exclusive). But if you're looking for a read meshed with pedantic observations of historically sad times overlaid with a bunch of stats and data, this book you're reading will not leave you wholly satisfied. Yes, there is pertinent historical context throughout this book, but I make it a point to be pretty plainspoken so not to bore anyone to tears. Hence, there are no scholarly anecdotes of famed Harvard professor Henry Louis Gates Jr. and me pontificating on the economic impacts of post-Reconstruction policy while donning ascots, smoking a classic pipe, and toasting our own intellect while sipping twenty-year-old brandy (but I am *totally* up for that if anyone wants to put me and the professor in a fancy library den lined with the world's best books). More importantly, the chattering class that has populated the cable news airwaves for so long can frequently make politics and policy hard to consume, as their perspectives are not often reflective of the diversity that drives democracy. So I set out to make this content, laced with my own personal stories, digestible and compelling for every reader.

Moreover, I am a subject matter expert. I've spent more than twenty years working in and around media and politics (please give an obligatory gasp here, followed by shocked sentiments of *"Twenty years? No way. Did she start when she was only five?"* Thank you.). I've covered presidential elections and administrations, Congress, and Capitol Hill for outlets like CNN and BET Networks. I founded and ran my own news platform. I've worked on numerous campaigns at the local, state, and federal levels. I've been a Black voter for more than two decades and I have had a Black voice for more than four decades. Hence, I'll not be entertaining too many questions about my bonafides. Perhaps it's challenging for those who have bought into the fairy tale that narrators penned about America to confront the actual truths about this country. When haters, under the guise of historians, began attempting to discredit the *New York Times*'s Nikole Hannah-Jones's the 1619 Project—a historical analysis of how the enslavement of Black Americans shaped US political, social, and economic institutions—it was disheartening. But it was not surprising. Some want to tell *our* story through their lens. Perhaps that's why I have seen more all-white panels discussing Black voters plastered across my TV screen than I care to remember. I'll get to that later. But first, I'd like to take you behind the scenes. Let's start with my journey in cable news . . .

ONE

How I Got Here

1

For Colored Girls Who Consider Journalism When Twitter Is Not Enough

Is it too hard to emphatically say Donald Trump was a racist businessman who became a racist reality TV star whom a little less than half the American voters elected as president? Apparently so. For a long time, much of the media found it impossible to declare. In fact, some still do. Perhaps this is why, for some Black people, fake Twitter and Facebook accounts became a more reliable source for news and information than actual "reputable" news outlets during the 2016 general election. The efforts of the Kremlin-sponsored Internet Russia Agency to manipulate Black voters depended heavily on its ability to feign consciousness and call out obvious racism in a way traditional media had not. Racism has so many different shapes and forms, purveyors and tentacles, and has such a huge impact on America. But for too long in the mainstream media, it has been something that dare not be spoken about in an authentic and honest way.

When this type of disconnect impacts politics, the consequences cast a large and dark shadow. Too many would describe Trump's outright racist remarks as having "racial undertones." If someone can tell me the difference between a person who makes comments with racial undertones and someone who is an outright racist, I'll send you a free copy of this book. If you take all of Trump's actions and rhetoric together, the pattern is clear: he is, by definition, a white supremacist. It is not just political opportunism but an ideology ingrained in his character. Yet for too long, the superfluous question that had been asked and answered was continually posed by the media: Is Donald Trump a racist? When Anderson Cooper posed this question to New York Democratic congresswoman Alexandria Ocasio-Cortez in a 2018 *60 Minutes* interview, she responded (as most seeing, thinking people would): "Yeah, no question." For some reason, this was the highlight of the interview. The next day, her response made headlines. It was discussed on cable news outlets as though she had said something controversial. Was this really breaking news? No. Still, most of the headlines responding to the interview keyed in on her stating the obvious.

But who cares about speaking a declarative truth when the debate itself is driving badly needed ratings for broadcasters. Just a few years prior, the networks had to figure out how to maintain a loyal audience. With hundreds of channels and curated and tailored information at your fingertips, there's a constant battle for the TV viewer's attention. In 2014, cable news was struggling to find its footing in this shifting media landscape, where its audience was not matching up with revenue trends. Cable news had peaked as a medium around the 2008 presidential election, and the business model was uncertain.

All three major news channels—Fox, CNN, and MSNBC—had seen their audiences decline. According to Pew Research, MSNBC's prime-time viewership had declined 4 percent in 2013, and 17 percent during the day. The Fox News channel had lost 5 percent of its daytime audience. And CNN's prime-time audience had declined by a quarter, to a median viewership of 495,000, according to Nielsen Media Research data. The total prime-time audience for the three channels combined hovered at the time at around 2.8 million viewers. That amounted to a fraction of the more than 20 million people who were tuning in to one of the three commercial news broadcasts airing on NBC, ABC, and CBS each night.

But with Trump, the cable news networks started enjoying great gains in viewership thanks to lapses in judgment by people who treated intolerance, xenophobia, and misogyny as though it were something to be laughed at and not taken literally or seriously. Former Democratic Minnesota congressman Keith Ellison took it seriously. He appeared on ABC's *This Week with George Stephanopoulos* in July 2015, was the sole Black person on the panel, and was practically laughed off the set when he said this: "All I want to say is that anybody from the Democratic side of the fence who's terrified at the possibility of President Trump better vote, better get active, better get involved, because this man has got some momentum. And we better be ready for the fact that he might be leading the Republican ticket." You can hear his fellow panelists—including the *New York Times*'s Maggie Haberman, ABC News political director Matthew Dowd, and Republican strategist Ana Navarro—laughing at his assertion. "I'm telling you," Ellison says, over the audible chuckling, "stranger things have happened." Stranger things indeed.

Did these members of the chattering class think it was impossible for a white supremacist to occupy the Oval Office? There have been far more bigots in the White House than there have been unprejudiced presidents. Surely they must be aware. Never before heard tapes were released in 2019 of then governor of California Ronald Reagan referring to African delegates to the United Nations in 1971 as "monkeys." He said to then president Richard Nixon, "To see those, those monkeys from those African countries—damn them, they're still uncomfortable wearing shoes!" He didn't say "shithole" countries, but the sentiment is certainly the same—these presidents viewed Black people as inferior and governed accordingly, shaping policy that directly and negatively impacted the Black community. During his first presidential campaign in 1976, Reagan dubbed a Black Chicago woman a "welfare queen," to perpetuate a vile stereotype of Black women living large on taxpayers' dollars. He then used her story to attack housing benefits, aid to children in poverty, and food stamp programs. Certainly the panelists were alive during his administration. Had they forgotten?

It wasn't such a surprise to most Black people that Donald Trump struck an old, reliable chord with many white Americans. But it was disturbing to see his narcissistic antics proving to be a huge boon for the networks. The outlets managed to reverse the trend of audiences tuning them out by casting a reality TV star in the starring role of their daytime and prime-time programming. Fox News, in 2016, led the pack, averaging 2.4 million viewers in prime time. But CNN narrowed the gap with one of its best years, averaging 1.3 million viewers during its prime-time slots. And MSNBC made an enormous leap from where it had been only a year prior, seeing the highest percentage gains of all the three news networks, averaging 1.1 million viewers.

The twenty-four-hour news cycle had found a star, and it was failing the American people. Bigly. Many cable news viewers were older white men—the median viewer age is sixty-six for Fox News, sixty-one for CNN, and sixty-three for MSNBC, according to Nielsen. And the networks' programming reflected the demographic. Given the type of campaign Trump was running, discussions about race frequently blanketed the networks. Diverse voices, however, did not—on or off screen.

There were endless reports of police brutality. The water crisis in the predominantly Black city of Flint, Michigan, had reached a boiling point. Immigration was also a hot-button issue. Trump, as the Republican nominee, offered up a slew of incendiary racist rhetoric. And when voters turned to cable news to unpack these issues, they were greeted by predominantly white analysts, pundits, and hosts. The TV screen did not reflect America. And there was little to subsidize the opinion alcove, since people were forgoing newspapers. Only two in ten US adults were getting news from print newspapers during the 2016 election cycle.

Broadcast was pulling out all the tricks. The "Breaking News" banner became a permanent fixture across television screens. Is all news breaking? No. And the most sporadic TV news viewers learned that the banner is there simply to attract and hold their attention and not actually to alert them to a worthy occurrence of national importance. Once viewers recognize that this marketing tool is deceitful, it dilutes how viewers respond to actual breaking news. Also, you no longer need to be a subject matter expert, an elected official, or an accredited journalist to earn a steady spot on the screen. You need only be someone who has a strong opinion, a modicum of credibility (to the mostly white people in charge), and the ability to drive ratings. Cable news is now where pundits hold court.

In the 2017 movie *I, Tonya*, Bobby Cannavale plays a reporter for the tabloid show *Hard Copy*. In the movie his character says, "[We were] a pretty crappy show that all the legitimate news outlets looked down on . . . and then became."

The new "news" reality TV. That's another good way to describe the industry as it started to alter. Argument took precedence over information. The cable news outlets loved Trump-inspired division so much that he didn't have to personally appear on screen to bring the regular news cycle to a screeching halt. In March 2016, all three networks spent thirty minutes broadcasting live footage of an empty podium bearing Trump's campaign sign ahead of a planned press conference, effectively giving him and his campaign a free thirty-minute commercial. And it wasn't like the networks were airing actual commercials. The news channels so loved the shock value of what happened at MAGA rallies that they covered the events live with no commercial breaks.

Over the course of the campaign, Trump earned close to $2 billion worth of free media attention. But not even the circus

Trump brought to town was enough to carry an entire story line for months at a time. The cable outlets needed a cast of supporting characters, because like any salacious reality TV show, the "news" circuit desperately depended on conflict. CNN seemed to find that formula with its multiple boxes of commentators squeezed onto one screen. Enter the "don't come for me" pundits on the left versus the "Trump is my God" sycophants on the right. Who's the anchor? Not important. Most were essentially assuming the role of the boxing announcer Michael Buffer, sitting atop a breaking news banner yelling, "Let's get ready to ruumbbbblllle."

Viewers didn't necessarily learn anything, but . . . were you not entertained?? Some talking heads were so offensive that CNN had to fire multiple Trump-supporting commentators. In 2017, Jeffrey Lord was fired for tweeting "Sieg Heil." And Ed Martin, a Trump-supporting talk show host, was fired for calling other CNN contributors "Black racists," and "rabid feminists." This was the intellectual discourse meant to inform the public? Not really. But still, many Black people were hungry to see this type of racist rhetoric swatted down in a public and humiliating way.

Enter Angela Rye. She burst on the scene delivering unapologetic readings of right-wing MAGA supporters and quickly became a household name and cable news favorite. She was part of the diversity the outlets so desperately needed. At the time she began doing television, she had recently completed her time as executive director of the Congressional Black Caucus (CBC). This accomplished lawyer had also previously held numerous positions on Capitol Hill and ran a political strategy firm. Yet she sometimes appeared on panels opposite people whose political experience paled in comparison to hers. She once appeared opposite shoplifter-turned-political-consultant Katrina Pierson, who falsely accused Rye of running the CBC "into the ground." Angela also battled on air with former congressman Joe Walsh, who said that President Obama was held to a lower standard because he was Black, to which Rye responded, "I'm not interested in having a dialogue with someone like Joe, who has demonstrated a propensity towards bigotry." Understandably, having to dignify such racist remarks takes a toll on the spirit. "I cried after," Rye told me. "I cry after a lot of segments because I can't believe this is where we are. At least these bastards used to hide their truth. Now, they are just bold with it, and it's disheartening."

But she has also received backlash from some in the Black community who found her sometimes brash delivery to be off-putting and occasionally disingenuous. I must say, Angela is a dear friend whom I have known for years, and I can personally attest that she is the same on air as she is off air. She is passionately driven by a mission bigger than herself. She is laser focused on uplifting Black people and goes out of her way to offer tangible support to those around her and the community at large. She once told me that being on cable was not enough. "If I leave this earth and all I did was be a voice for some Black people on TV, then I have failed. I'm not trying to build a brand; I am trying to build up my community." Which is why it may be all the more hurtful when the criticism comes *from* the community. I asked her how being on the receiving end of such harsh critiques impacts her. "Black people aren't monolithic. Many say I speak for them. I suppose this is the way of the critics saying I do *not* speak for them. It's all good," she replied. "I wish there was a constructive way to go about it, though. I never want my words to be the source of pain for the culture. So to say that some of this criticism has hurt me deeply is an understatement."

The argument can be made that Black voters, particularly younger Black voters, saw someone who not only looked like them but spoke their thoughts—and they tuned in. CNN saw a 5 percent uptick in their Black viewers in 2017. A more diverse slate of talking heads who were able to call out racist attacks in real time—something many anchors and reporters seemed incapable of doing—arguably brought hundreds of thousands of people into the political process. I can't say for certain that anyone learned anything from these types of segments. After all, when you have eight people on a screen all talking over each other in heated disagreement, the intellectual exchange gets lost. But seeing people the audience found familiar certainly piqued their interest. One could also make the argument that having more relatable personalities across screens makes politics more palatable to younger generations as well, and may have impacted turnout during the 2018 midterm elections. Generation Z, mil-lennials, and Generation X accounted for a narrow majority of midterm voters. The three younger generations—those ages eighteen to fifty-three in 2018—reported casting 62.2 million votes, compared with 60.1 million cast by baby boomers and older generations. So who says young people don't vote?

They certainly do on social media. In fact, it appeared that was the intention of these segments: put an obnoxious person on air to square off with a combative guest, and voilà, you've got a viral hit on your hands. After each of these explosive cable news segments, it was time for the judges (the audience) to cheer their champion and jeer the opponent. Break out the smartphones. "Like" the obnoxious clips of shouting matches and retweet all the comments you agree with. Tag the opposition in an angry diatribe and eagerly await your favorite pundit to tweet some epic clapback to some right-wing Trump supporter. Find everyone's public page on Facebook so you can like it or move over to Instagram so you can follow. Google the cable TV personalities to learn about their personal lives, whom they're dating, if they're married, study their fashion choices, and so on. But ask the masses to intelligently reiterate the point either side was making? (Assuming there was such a thing.) Nope. Name the issue they were discussing? Verify any data they quoted? Nah. No time for that. If audiences of color were becoming more engaged, cable news was certainly not making them more politically astute.

Long before I ever appeared in front of the camera, I spent a significant part of my career behind the camera. In fact, it's where I got my start. I cut my teeth on news gathering at CNN. I remember the day I was hired as an entry-level video journalist, in the year 2000 at the world headquarters in Atlanta—it was when the network was celebrating its twentieth anniversary of being the pioneer of the twenty-four-hour cable news. And I got to be part of this innovative media giant. I'd work whatever shift they gave me without complaint. Work weekends? You bet! Work on Christmas Day? Sign me up! Overnights? No problem! I just wanted to be there. I could not believe that

CNN was doing me the favor of letting me work twelve-hour days and giving me the honor of getting yelled at in the control room—just like a real-life scene in the 1987 classic *Broadcast News*. The best part? They were paying me a whopping $22K a year to do what I would have done for free. Mama, I made it.

But I soon discovered there was no welcome mat for me as I entered this world. The journey to get an entry-level slot at CNN had already been filled with pitfalls, and I was surrounded by people who had nothing but launching pads in their youth. I had been on my own since I was sixteen years old. By hook or by crook, I found my way to college—Clark Atlanta University, an HBCU (Historically Black Colleges and Universities) in Georgia. But after just a couple of years, I ran out of money and had no idea how to pay my tuition. I had to leave. I had been on the homecoming court, I befriended my professors, I forwent parties for symposiums. Being ejected from the nourishing environment of a historically Black college was devastating. I spent months feeling like a colossal failure destined to live below the poverty line. I was watching my peers soar while I ate ramen noodles and read newspapers on the floor of my apartment because I couldn't afford furniture. What the hell was I going to do now?

Hell-bent on not letting my circumstances define me, I became desperate. But more than being desperate, I was hungry. I began reciting daily my own version of the famous Henry van Dyke quote, "Some succeed because they are destined, I will succeed because I am determined." I wrote it on my bathroom mirror in lipstick and would read it out loud, blinking away tears. I began stalking the program director at the biggest urban radio station in Atlanta, V-103 WVEE. I sat in the lobby every day for a week before he agreed to see me. He gave me fifteen minutes, which I used to beg and plead that he give me a shot at news. I had my résumé, printed copies of writing samples, my reel, references, anything he could possibly want. Perhaps out of pity or to get rid of me, he finally agreed to let me fill in doing the morning news slots on their flagship program—the Frank Ski morning show.

The experience of working in radio and being responsible for content made me feel like an Oprah-in-the-making. I took this responsibility very, very seriously. Though no one was ever going to actually lay eyes on me in radio, I wore the nicest of the cheap threads I had in my closet. Nearly one hundred thousand people were tuning in to hear the information I brought them. This was serious business. I had to treat this opportunity as though my life depended on it—because, at that time, it felt like it did. The hosts seemed to love me. It was working well for a few months. I don't remember what freelance rate they were paying me, but it wasn't a lot. Yet it was still unbelievable to me that they were paying me at all. So when I met an African American CNN executive while attending a college football game, I was grateful that I had made a practice of carrying copies of my résumé with me—this was before everyone was on email (news flash: I'm actually not a millennial, I just play one on TV—don't tell anybody!). His name was Gerald Walker. He told me he would speak to someone about me and that all he could do was guarantee an interview—not a job. Fine with me! An interview was all I needed.

To my surprise, a few weeks after my interview, I was hired. I was getting paid an actual salary to be a journalist. I had pursued and studied this field since I was fourteen years old. I had written columns for the *Cleveland Plain Dealer*. I had learned how to produce packages in the National Association of Black Journalists' Urban Journalism workshops. I had seen every single episode of the *Oprah Winfrey Show* and tried to make my voice more of an alto, since her timbre came off as more serious than my soprano. I had shaped my formative years around this industry, and it was finally happening—I was entering my field in the most amazing way. But trust me when I tell you that if your goal in life is to get wealthy, working in cable news is not the move. When I made my entrée to this new space, my idea of money was very different. And it felt like everyone had it except for me. My entry-level colleagues all lived in cute downtown apartments not far from CNN. I lived on the south side of Atlanta, known as the SWATS, in one of the few apartments that would accept me, after I slipped the application manager $100 to ignore those unpaid college credit-card blemishes on my credit report. The blessings of growing up not privileged: I knew how to clear the path when I needed to. Amen?

My colleagues all looked like they belonged. They rocked designer handbags and wore shoes I had never heard of. I rotated three of my best bargain-basement suits, which I used to wear with sneakers and pretend I left my designer heels at home—very *Working Girl.* That was believable, right? I couldn't even afford a cheap pair of heels at that time. My colleagues all drove cars. New ones. I enjoyed the comfort of Atlanta's public transportation system. And I could barely afford that. In fact, there were many days when I would feign putting money in the machine on the subway system and then tell the person on duty that I had paid the fee but the entry doors never opened. Hello? I'm in my *Working Girl* attire with my CNN badge conspicuously displayed. Obviously I'm telling the truth.

My journey was definitely weighted with struggle. But I was not unique. The path to working in TV news is narrow for most Black people. The median income for a white family is forty-one times higher than the median income for Black families and twenty-two times higher than the median income for Latino families, according to the Institute for Policy Studies. So how many young Black and brown people will pursue careers that are not the most lucrative (e.g., journalism), and who can then afford to live in one of the major cities where cable news networks operate? Not a lot. And for colored girls who consider journalism when Twitter is not enough, the pipeline to land an internship in news is limited. Most news internships don't pay. When it comes to the cable news outfits, they all operate in major cities—mainly DC and New York (with CNN's headquarters being in Atlanta). Finding an affordable place to live in these major cities can be tougher than landing the internship itself. For example, the costs of interning in DC (where I live) for a summer can total up to $10,000, when you consider an apartment, professional clothing, transportation, and food. Not too many Black and brown students can afford to work and get paid only in experience and not actual compensation. So for many students, an unpaid internship just isn't in the cards. And if young people of color can't afford to intern in cable news, they will never gain the valuable experience of learning how to navigate a newsroom. The genre should be welcoming the next generation of passionate reporters, producers, editors, and news gatherers. But sadly, for so many students of color who are struggling just to stay in school, when it comes to opportunities—if it don't make dollars, then it don't make sense. I certainly didn't have the financial support my colleagues seemed to have. I didn't have parents subsidizing me. Both my parents, who were never married to each other, were addicts. My father passed away from alcoholism when I was eleven. My single mother was battling alcoholism three states away (she's sober today!). I didn't have a single family member in a position to help me. There were no wealthy aunts or uncles or cousins to swoop in and save the day. And no one in the newsroom was interested in my circumstances. This was my job, the beginnings of my career. And my superiors—the people who would be making decisions over who advanced to the next level—were not counselors there to guide me or mentors there to sympathize. I had to figure it out. Alone. And I did.

I showed up to work every day, even when I was not supposed to be there, so I could observe and learn. Also, I couldn't afford cable. So I couldn't talk about the great segment everyone had seen in the comfort of their homes—I had to be there to read all the free papers and know everything that was happening by watching one of the hundred TV screens in the newsroom. I also couldn't afford a computer. At CNN headquarters, there were plenty. I could spend hours reading the wires. Back then, I focused on global news. Had you asked me the latest with then UK prime minister Tony Blair or the doings of Palestinian president Yasser Arafat, I would have had a ready answer. Then there were the times that I had to hide the sheer panic on my face when they assigned me the 5 a.m. shift. Atlanta's public transportation system didn't *start* running until 5 a.m.,

which meant the earliest I could get to work would be 5:30 or
5:45 a.m., after I took the earliest bus to catch the earliest train.
This was not going to work. Taking a cab every day was something I definitely couldn't afford. On these days, I had to get a
ride however I could, frequently with a neighbor who worked
overnights. He would drop me off at midnight, a full five hours
before my shift, and I would find quiet edit bays to crash in
until it was time to work.

There were so many obstacles at that time, but my cockeyed
enthusiasm helped me persevere. After a year of working in the
world headquarters, I was offered a job in the nation's capital
to cover politics as an editorial assistant. It felt like a dream.
Which is good, because when I arrived in Washington, DC,
that was all I had—a suitcase and a dream. I had long wanted
to be the brown Murphy Brown, and I was finally here. This was
going to be epic. I had only been to the Beltway once, and that
was for the interview. And then it took everything for me not
to stroll into the middle of traffic and do my Mary Tyler Moore
spin and hat toss directly in front of the Capitol Building. Seeing where Congress convenes, the Lincoln Memorial, and all
the people who made up Chocolate City (back when DC *was*
Chocolate City)—it felt like I was on the set of a movie. It was
as if I had floated out of my body and was watching myself on
screen, living the life I imagined. But of course, like any good
movie, there was lots of drama.

I may have been happy to join CNN's DC bureau, but not
everyone was happy to have me. I didn't have the right pedigree
for the chattering class. I wasn't the daughter of some senator. I
wasn't dating some congressman. My parents weren't longtime

friends with some ambassador, who had made a few calls to get me the job. And everyone around me seemed to know things I didn't. I worked for Bob Novak, the late syndicated columnist and conservative political commentator. He hosted the show *Capital Gang* for the network and was known as the Prince of Darkness. With me, he frequently lived up to that name. As a young person in the newsroom, I desperately wanted him to like me. I made small talk, I referenced some of his columns, I asked questions. Nothing worked. But I considered it a feat when at least he seemed to stop hating me. I also got to work with the very pleasant Wolf Blitzer, who at that time was hosting the Sunday morning political show. He was never without a compliment or some words of encouragement for me as he passed me in the halls. But I still felt way out of my league. The brown Murphy Brown was not winning over her colleagues. And the worst part was that some of the blame was mine: I made many mistakes. I was used to reading a paper or two every morning. My colleagues were all reading five or six. I'd get control rooms confused and give the wrong scripts to the wrong directors. I'd address congressmen as "senator" or ambassadors as "secretary." I hadn't yet been schooled on all the proper greetings and titles. And I could never keep up with all the acronyms my colleagues spouted all day (partially because they had all been in DC politics since before I was born). I had to ask for clarity as though I had just forgotten. Ummm . . . the DOJ? Ah, yes! The Department of Justice. Got it. DOD? Oh, of course. The Defense Department. Duh, I knew that—just a brain freeze, guys.

I was often left out of office conversations. After all, I had not yet made it onto the invite lists to the fancy weekend parties at embassies and museums. I rarely got their inside jokes, which one could only understand via the osmosis of decades of Beltway politics. I was twenty-two years old, so I did not yet have the wisdom time provides. I was made to feel like an imbecile by my superiors and colleagues if I didn't know the detailed policy points of some famed event that occurred in the 1920s. *Obviously, Tiffany, it was Woodrow Wilson who created the Federal Reserve.* Then they would look at each other in bewilderment as if to wonder how I could possibly not know that. How indeed. I wondered how well they would fare if I tried to engage them in a conversation that was relevant to me regarding the early part of the last century. Would they be able to offer any thoughts about the dichotomy between the approaches to social and economic progress that W. E. B. Du Bois and Booker T. Washington advocated for Black people? Because had they asked me about that, I might have had a thoughtful answer. But their American history always seemed to matter more than mine—even though both were parts of our shared history.

My colleagues would also have water cooler conversations and laugh boisterously together as they dropped movie lines from what they considered classics: *American Pie*, *Revenge of the Nerds*, *Something about Mary*, and so on. Sure, I had seen these movies, but they weren't classics for me. I wondered if they would laugh if I dropped a classic line from *Do the Right Thing*, *Coming to America*, or *Friday*. Yet these were the people making important editorial decisions, scripting the anchors (mostly white) with banter, and discussing what they thought mattered in Congress, the White House, and the judiciary. How had their experiences shaped their view of what was newsworthy and what was not?

One day in 2002, in our *Capital Gang* news meeting, one of the cohosts of the show was laughing about a story he wanted to discuss. It was President Bill Clinton being inducted into the Arkansas Black Hall of Fame as an honorary member. He found this story hilarious. Incidentally, I did too, but for different reasons. Clinton was to be the first person who was not African American to be recognized in the hall's history. The entire team was laughing, but I stayed out of it and let them have their fun. Then one of the show's cohosts, barely able to contain his laughter, said to everyone, "Wait, guys, here's the best part. He's being inducted with some singer no one has ever heard of. I'll give anyone in this building $100 if they can name one song by this guy . . . what's his name? Al Green." They all joined in the outbursts of uncontrollable laughter. "Who?" they asked through their chuckling.

Are you guys kidding? Al Green is a legend who recorded a series of hits in the early 1970s that are still in heavy rotation today. But I stayed silent and tried to contain an almost involuntary eye roll. One of the members of the group, noting my silence, tried to bring me into the conversation: "Come on, Tiffany. Have you ever heard of Al Green?" I replied, "No, and I've got to go cut these sound bites." I don't know why I didn't answer truthfully. Maybe because I didn't want to be responsible for turning a fun moment awkward for all the white people laughing. Maybe because I didn't think I needed to justify an R&B legend's significance to any of them. Maybe because at the tender age of twenty-two, I was already tired of being "the only one." Either way, I abstained from the chatter. The truth was—I could and *would* eventually learn all of the things they knew. But those poor souls had gone their entire lives without ever having enjoyed the smooth, melodic sounds of *Let's Stay Together*, *For the Good Times*, *Simply Beautiful*, or *Love and Happiness*. And they had the brass to laugh at the idea of a man whose falsetto could damn-near impregnate people? At that moment, I was grateful to be me and not them.

Other people made mistakes as well. But to me, young white girls in the newsroom (and pretty much everywhere) seem to always be forgiven and guided in a way that I never was. They're looked at with familiarity or lust by many of the white male executives in charge. Right, Roger Ailes? I got a small bump in pay when I moved to DC, but the cost of living was higher, so I was still in the same financial situation when I left Atlanta. I worked on all the weekend shows, so I had to be at work at 6 a.m. on Sundays to read the morning papers and flag anything of interest for the producers. Just one small problem for me: DC's public transportation did not start running until six on Sunday mornings. On these days, I had to take a cab into work. It cost $17 with no tip, because I could not afford to live on Capitol Hill or in Georgetown; I was living in a studio apartment in Silver Spring, Maryland, about twenty minutes outside the city (with no traffic). And $17 was nearly $70 a month, which was a lot on a limited budget.

There was a young white woman on the team who happened to be very pretty. I liked her a lot. Everyone did. She was bubbly, friendly, always came in with a smile. She was having a big birthday party one Saturday night and had invited everyone. But she too had to be at work the next day at 6 a.m. I overheard her asking the supervising producer, a white man, if she could come in late the next day. After all, it was her birthday party, and she couldn't be expected to get to work on time after celebrating all night. She promised she would read in from home and be ready to go by the time she got there. He happily obliged. Huh? That was an option? I mean, I wasn't having a birthday party or anything, but it sure would be nice to not have to count pennies at the grocery store. I agonized for a week about asking the supervising producer the same thing. How would I broach the subject? What would I say? Finally, after practicing, I was ready.

"Hey Bob, do you have a sec?" I asked, trying to sound super casual. He replied, "Sure." I explained to him my financial situation and that because of where I lived the cabs cost $17 a week. Would it be possible, I asked, if I could read the papers from home on some Sundays. Not every Sunday, but maybe just some Sundays? Just so I could save a few bucks. He frowned and then sternly said, "Let me get back to you." Gulp. Now before I tell you what happened next, let me say that looking back, this may have been an inappropriate request. Getting to and from work is the employee's responsibility, and it's not the supervisor's role to figure it out for them. But after he had been so cheerfully accommodating to my colleague, I had hoped for the best. She was asking for one Sunday. I was asking for it to be a semiregular thing. He liked her. He tolerated me. She was part of the newsroom clique. I was an outsider. Still, what happened next seems to be a bit of an overreaction in hindsight.

The supervising producer had gone to the executive in charge of *all* weekend shows—someone three tiers his senior and five tiers my senior, a white woman who also tolerated me. She called me into her office and shut the door. She proceeded to tell me how inappropriate it was for me to go to my superior and complain about money problems, expressed regret in hiring me, and told me that I, not anyone else, needed to be responsible for how I show up to work. She then took $20 from her purse and tossed it across her desk at me and said, "Use this for this weekend and show up on time. Next week, you're on your own." I sat across from her humiliated. I felt my throat closing and the tears burning my eyes. Before one tear could fall, I replied, "I'm really sorry I ever mentioned anything. It won't happen again. I will be here Sunday on time. I don't need your $20. Thanks anyway." I left and had to walk right by the supervising producer, who conveniently avoided eye contact. I cried on the train ride home. And that Sunday, I took a cab into work that I paid for myself. I showed up and did my job surrounded by people who all seemed to view me as the problem child of the newsroom. It was a lonely place to be. But given all that people who looked like me had endured, surely this would not be the thing to break me. Still, the lack of diversity in news at that time was paralyzing because it also meant a lack of understanding. Later, Tracey Webb, a Black producer, joined the team. She was more senior than me and lived in my neighborhood, and out of the blue one day, she asked if I would like to ride to work with her on Sundays. I did. She was a godsend.

If people of color don't feel welcome in the newsroom, they leave. The pipelines created should be driving diverse voices toward this space where they are so desperately needed, not away. During my time at CNN, I frequently felt like the news business was not for me. I eventually left and explored other career paths in media. I thought about other ways I could tell stories.

I stopped watching cable news programming for a few months before being sucked back in. After seeing how the sausage was made, every time I viewed the content, I thought about the perspective of the editorial decision makers, most of whom did not look like me. In the early 2000s, white men were laughing at the concept of this "unknown" singer Al Green sharing a stage with President Bill Clinton. What are they laughing about today? What's lost on them today?

Having a newsroom reflective of society also impacts coverage. In May 2017, Trump tweeted that there might be recordings of his private conversations with former FBI director James Comey, whom he had just fired. It was a clear attempt to caution Comey against "leaking to the press." It sent mainstream outlets into a tizzy over whether he recorded conversations. That same day, then attorney general Jeff Sessions rescinded an Obama-era rule telling federal prosecutors to avoid charges for low-level drug offenders that could trigger lengthy mandatory minimums. Just like that, mandatory minimums were back. Where was the breaking news banner for this devastating policy that would overly impact Black and brown people? I don't recall any network giving it due coverage. Newsroom diversity literally impacts democracy.

MSNBC, where I appear frequently, has been criticized for relegating most of their on-air talent of color to the weekends. Actually, all the networks share this criticism, but MSNBC has a loyal audience of Black viewers. In 2017, the network drew more African American viewers during prime time than any other cable network. And that's not just any other cable *news* network—this included every channel on cable. Between 2016 and 2017, the network saw a 50 percent increase in Black viewership. But most of their senior executives do not look like their viewers. This showed in an egregious misstep regarding a memo about Iowa Republican congressman Steve King, who has a well-documented history of promoting white supremacy and making racist comments. In 2005, he sued the Iowa secretary of state for posting voting information on an official website in Spanish, Laotian, Bosnian, and Vietnamese. In 2013, he opposed legal status for Dreamers, undocumented youth who had been brought into the country as children, saying, "For everyone who's a valedictorian, there's another hundred out there that weigh 130 pounds and they've got calves the size of cantaloupes because they're hauling 75 pounds of marijuana across the desert." He said US culture cannot be restored "with somebody else's babies" and advocated for "an America that's just so homogeneous that we look a lot the same." He has called Western civilization a "superior culture" and displays a Confederate flag on his desk in DC. He also met with leaders of the far-right Freedom Party in Austria, which had been founded in the 1950s by former Nazis. We could fill many pages detailing King's racism in both rhetoric and policy. But here's where it intersects the lack of diversity at the network level.

In a January 2019 interview with the *New York Times*, King asked, "White nationalist, white supremacist, Western civilization—how did that language become offensive?" This is inarguably a racist statement. Yet, following the incident, the standards department at NBC News sent an email to staffers instructing them not to directly refer to King's comments as "racist." Two staffers leaked the email to *HuffPost*. "Be careful to avoid characterizing [King's] remarks as racist. It is ok to attribute to others as in 'what many are calling racist' or something like that." Huh? What in the name of Spike Lee must one do to actually be considered racist? Talk about how wearing blackface is okay if it's Halloween? That seemed to do the trick when the network finally fired former Fox News anchor Megyn Kelly, who had previously said Jesus and Santa were white, and that Sandra Bland would be alive if she had complied with police orders. Unfortunately, by the time network executives fired the well-documented racist they had hired, journalist Tamron Hall—the first Black woman to ever host *The Today Show* and a household favorite—had already been edged out to make room for no-melanin Megyn. But back to King. Only after reports emerged about the email did NBC News reverse course and send another email to staffers revising their previous guidance. "It is fair to characterize King's comments as 'racist,' and point out that he has a history of racist comments, and the context can be shared that others hold that view as well," read the second email. The policy would have likely stood were it not for the leak and the outcry that followed.

At the end of 2019, CBS News shared a bizarre video of Trump supporters in Hershey, Pennsylvania, threatening that they would angrily take up arms and ignite widespread violence rivaling the Civil War if the president were removed from office. A white man in a gray Trump shirt and a red MAGA hat leaned in toward the camera and ominously repeated, "He's not gonna be removed" three times before saying, "My .357 Magnum is comfortable with that" when an unseen reporter asked if he was sure of Trump's job security. The next Trump supporter, decked out in a white Trump 2020 cowboy hat, said he would expect to see "physical violence" should the Senate remove the president. "I think it would become a second Civil War," he warned. Aside from this video being batshit crazy, the most grotesque thing about it is that CBS presented it as though these were just regular sound bites from people on opposite sides of the impeachment debate. They tweeted out the video with this text, "Trump supporters at the president's rally in Hershey, Pennsylvania, tell CBS News their thoughts on impeachment and the possibility of his removal: 'It would become the second Civil War.'" Their thoughts?? I wonder, if these were a group of Latino men threatening violence if Congress voted to overturn DACA, would it simply be people "sharing their thoughts"? If this were a group of Black men threatening mass violence against the country if sentencing reform didn't become a legislative priority, would they just be sharing "their thoughts"? This narrow, myopic view draped in white privilege abandons viewers and blankets all coverage in questions of editorial judgment. And once you lose the trust of viewers, it's hard to get them back.

CNN's morning show has a segment called "The Good Stuff," where they highlight one positive human-interest story. At the height of news coverage about police brutality in the Black community in 2016, the show aired a video of an interaction between a group of Black men and a police officer who happened to be white. The Black men were groomsmen in a wedding the next day and had planned to do a song-and-dance routine. They were in the parking lot of a hotel rehearsing when someone called the police to complain. The police showed up to investigate. The guys explained what they were doing and showed the cop the entire routine. The cop appeared to have enjoyed his free show and left without incident. Annnnd scene. Coming out of the segment, the anchors seemed positively wrapped in all the warm and fuzzies. Meanwhile, I was watching this foolishness and felt enraged. This was the good stuff?? Good news, Black people! All you have to do to not get shot by police is perform your impromptu *Five Heartbeats* song-and-dance routine for them and, voilà, you get to make it home safely that night? I am certain the hosts did not intend this segment to be offensive. The optics were just painfully lost on them. What if there had been a Black person hosting the morning show during this segment? Or a Black producer who was part of the editorial discussion? Perhaps they might have been able to highlight how unintentionally insulting it was to celebrate a cop for managing to not shoot the group of Black men rehearsing for a wedding only after they doo-wopped and diddied for the officer.

A few weeks later, the same segment highlighted a young Black man with a darker complexion and locks, wearing baggy clothes, who had been captured on camera entering a convenience store that happened to be closed. Apparently the owners had forgotten to lock up shop. The customer was unaware and can be seen perusing the store, grabbing a beverage, and standing at the counter patiently waiting for someone to charge his purchase. When, after a few minutes, no one came, the young man pulled cash out of his pocket and left it on the counter and then exited the store. Once again, this was the good stuff? Perhaps the producers and hosts who thought this would make for

a great feel-good segment all had the best of intentions. However, that's not how it was received. What exactly was the good
stuff here? Surprise! Check out this young Black guy not steal-
ing! Who knew? By making this the good stuff, the unintended
insinuation is that the young man in the video is an anomaly—
one of the good Black kids who doesn't steal. I wondered, would
they have aired this video had it been a young white kid in the
suburbs in khakis and rocking a crew cut?

Even as newsrooms become more diverse, if people of color
are not in positions of power it can still feel like the white
power construct maintains editorial control. Only 7 percent
of newsroom employees are Black. And according to Pew Research, only 6 percent of news directors—who constitute the
leadership of such newsrooms—are Black, up from 2 percent
in 1995. I spoke with some current CNN employees of color
to find out how my former employer had changed in the years
since I had been an employee. Several told me about a morning editorial call with the senior decision makers to discuss
the news of the day. One story: During the summer of 2019
an African American man was waiting for his friend outside
an apartment building in San Francisco when a white man,
alongside his son, approached him asking to know the name of
the friend he was waiting for and accusing him of trespassing.
The encounter was captured on cell phone video. The white
man, since identified as Christopher Cukor, called the police
despite his young son begging him not to. Cukor defended
himself by saying he reacted based on his past history—he
said his father had been killed when he confronted a trespasser
outside his home. On CNN's morning call, a white executive

defended Cukor's decision to call the cops and advised that because of Cukor's unique history the network should come at the story a different way. Were there Black people on this call? Yes. Were there Black editorial decision makers on this call? No. Hence, the more junior people of color did not feel it was their place to speak up. There was no one to point out that this is frequently the excuse white people use when targeting Black people. In one fell swoop, this man's prejudiced behavior was dismissed and excused.

Those in charge of making editorial decisions still don't seem to see the glaring disrespect in some of their stories. In the fall of 2019, a CNN story highlighted a group of moderate Democratic white congresswomen as "leaders" on impeachment. The story was part of CNN's "Badass Women of Washington" series. I should note that this same series included Kellyanne Conway as a "Badass Woman of Washington," so hopefully that helps put the entire concept of this story in context. CNN credited the five white women lawmakers, who only switched their anti-impeachment views after new allegations in the Ukrainian scandal, as having "changed history by becoming unlikely leaders on impeachment." Huh? The story said nothing of the fact that many members of Congress who are people of color were among the first to push for impeachment. There was no flattering story about Representative Maxine Waters (D-CA), who was the first congressional Democratic woman to explicitly call for Donald Trump's impeachment in 2017. Nor was there any respectful mention of "the Squad": Democratic congresswomen Ilhan Omar (MN), Alexandria Ocasio-Cortez (NY), Rashida Tlaib (MI), and Ayanna Pressley (MA)—all of whom backed an impeachment inquiry long before these moderates. The story was painfully laced with tropes. CNN reporter Dana Bash actually asked the women, were they the "anti-squad"? That's when Representative Elissa Slotkin (D-NY) says, "None of us is ever going to get in a Twitter war with anyone else. If we have a concern with someone, we're going to go right up and talk to them about it, and we're not going to add unhelpful rhetoric to an already bad tone coming out of Washington." This was followed by Representative Abigail Spanberger (D-VA) saying, "I don't think any of us wants to be the loudest voice in the room. I just want to be the most effective." Seriously? No one heard this and saw the subtext of "loud" and "effective" being mutually exclusive, thus insinuating the women of color were loud and the white members of Congress were effective?

I won't waste too much time discussing Fox News. The obvious racist tropes echoed on that network are too many to enumerate in one book. From the 2008 presidential election season, when Fox News referred to Michelle Obama as "Obama's baby mama," to prime-time host Laura Ingraham's nightly monologues, Fox is a joke of a "news" channel. In 2009, former Fox News vice president turned Donald Trump campaigner Bill Shine said that if Barack Obama became president, Fox News would become the voice of the opposition. He was right. And their viewership has ballooned, far outperforming all other cable news networks. Going into the election, 19 percent of voters called Fox News their primary news source. CNN came in second at 8 percent. Fox is willing to put people of color on the air because too few established talking heads will appear on the channel. It's, in fact, the first cable news network on which I appeared, until I realized the shifty antics were never going to produce intelligent discourse.

The National Association of Black Journalists in 2019 began calling out CNN for its lack of Black representation in the organization's senior ranks of executive news managers and direct reports to CNN president Jeff Zucker. At that time, Zucker had no Black direct reports. There were also no Black executive producers, no Black vice presidents or senior vice presidents on the news side at CNN. The group explained that "the absence of blacks in news decision-making roles impacts the network's ability to provide balanced perspectives from one of the most influential and largest consumer groups in the nation— the black community." Soon after, this changed. The network named several African Americans to their c-suite and in the fall of 2019 named Emily Atkinson the first Black executive producer of Wolf Blitzer's *The Situation Room*.

The staffing of American news media has never reflected the diversity of the nation. Journalist Ida B. Wells covered lynchings in the 1890s that mainstream news outlets overlooked. It was African American news outlets like the *Pittsburgh Courier* that, after World War II, led the Double V Campaign, which aimed to achieve victory both in the war and in civil rights by pointing out the risks Black soldiers and civilians took fighting abroad, while being denied their rights as American citizens back home. During the civil rights movement era, with few Black journalists in mainstream newsrooms, newspapers and television networks struggled to cover the movement. No major Black voices were given a platform at any of the networks to help guide the discussion. In 1968, members of the

Kerner Commission—an advisory board formed by President Lyndon Johnson in response to a series of race riots—called out the role of a mostly white media in failing to cover the root cause of unrest. The report declared that "the journalistic profession has been shockingly backward in seeking out, hiring, training, and promoting Negroes." It added that "the press has too long basked in a white world, looking out of it, if at all, with white men's eyes and white perspective. That is no longer good enough. The painful process of readjustment that is required of the American news media must begin now." It called on news outlets across the country to diversify. But years later, not nearly enough progress has been made.

This is why niche media markets, like the Black press, have such value. I served in various positions in the BET News Department, briefly running the DC bureau. This role gave me a unique perspective on news gathering and the audience's proclivities when it came to news viewing. During my time there we covered the country's first Black president rather extensively, given that we didn't have a huge operating budget. During those first few years, we carried the president's State of the Union addresses, followed by impressive panelists including the *Washington Post*'s Jonathan Capehart; Republican strategist Sophia Nelson; the dean of the Black press columnists, the late George Curry; and others. But the viewers did not tune in in large numbers. Despite BET's coverage and unique perspective, its mostly Black audience would rather get its Obama news from mainstream news outlets. I surmise that for some, BET had let down older viewers with some of its programming. And others wanted their news validated by what a majority of Americans were watching. Instead of populating Black-owned media outlets, Black reporters, hosts, pundits, writers, and so on have clamored for the opportunity to appear on air in mainstream media. But they would find that space increasingly crowded and their voices routinely left out.

In a time of increasing racial hostility, Donald Trump has crystallized the term "fake news" by sowing distrust in media that does not praise him. But if, as the *Washington Post* declares, democracy dies in darkness, it also dies in whiteness. If a free press is to be the guardian of this representative government, then newsrooms should be representative of the country. And they still, blatantly, are not. More than three-quarters (77 percent) of all newsroom employees—those who work as reporters, editors, photographers, and videographers in the newspaper, broadcasting, and internet publishing industries— are white. About six in ten newsroom employees (61 percent) are men, making almost half (48 percent) of newsroom employees white men. In September 2018, the American Society of News Editors (ASNE) had to delay releasing the results of their 2018 Newsroom Employment Diversity Survey. Why? Because few outlets bothered to participate. Only 234 out of nearly 1,700 newspapers and digital media outlets submitted data. Extending the deadline helped a little, with responses rising from 234 to 293. The purpose of the ASNE diversity survey, which launched in 1978, is to document employment trends in US print and online publications and help newsrooms reflect the growing diversity of their audiences. The survey measures progress toward ASNE's goal of having the percentage of people of color working in newsrooms nationwide equal that of people of color and those underrepresented in the nation's population by 2025. In the survey responses, people of color represented 22.6 percent of the full-time workforce in US newsrooms among responding newspaper-digital organizations, compared to 16.5 percent in the 2017 survey. For the digital-only newsrooms, the figure was 25.6 percent, compared to 24.3 percent the previous year.

As diversity in the ranks of news outlets increases, so too will their consumers. And isn't that a good thing? News executives, who are overwhelmingly white and male, must learn to value the contributions of people of color—on air, in newsrooms, and in the corner office. According to Nielson, cable news viewers look like the men who run the networks. They too are overwhelmingly white, male, and middle-aged to senior citizens. If a majority of cable news viewers are white men over the age of sixty, surely these same execs must realize that their contemporaries are not immortal. There's a limit to their time driving viewership. When people of color don't see themselves, they tune out. Younger, more diverse audiences have begun to rely on social media for their news and information. A 2015 American Press Institute survey found that 88 percent of millennials say they get their news from Facebook—which we later learned is highly problematic. As the demographics of the country change, the media must catch up. Why actively ignore a growing influential market, which drives culture and conversation, by catering to a smaller, whiter sect of society. Punditry cannot be the only place news welcomes diverse perspectives—and that space is rife with challenges. I have never had any desire to consume reality TV. Watching outlets broadcast arguments that could once only be seen on Bravo's *Real Housewives of* (insert city here) has a shelf life. It must. Black people have too much on the line and, just like the rest of the country, deserve to be informed with thoughtful information and analysis that represents our unique perspectives. There were too few doing that in 2016. And much to my own surprise, I was about to join the fray. Twirl on that.

2

I'm Beat!

After being absolutely exhausted with the media landscape, I wanted to disrupt it. But how? I had always secretly envisioned myself running a multimedia platform but had no idea how to do it. During my time at BET, I tried to jump-start a weekly newsletter to engage the audience, but the company never really valued the idea. The news team already didn't bring in money for the company, and all my ideas cost money to execute. I spent some years going through an existential crisis. I had left the broadcast side of things, and I was incredibly unhappy. As a consumer, I felt ignored by the news. As a recovering news producer, I felt ejected from the business I had worked so hard to join. One night on my couch, reevaluating my life for the third time that week, I surveyed the scene. There was an obvious void. There was no outlet uniquely focused on the intersection of politics, policy, and people of color.

Certainly there were places Black folks could get news that

catered to our demographic. But many were often peppered with celebrity gossip and sensational clickbait. But I was surrounded by people who could intelligently discuss it all. We could debate the lyric abilities of Nas and Jay-Z with the same passion as criminal justice reform. We could talk about Michelle Obama's fashion in one breath and, in the next, debate her husband's place in the diaspora. We could speculate over who we thought might be "Becky with the good hair" in Beyoncé's album *Lemonade* and then discuss the footage of "barbecue Becky" harassing Black people having a cookout in a public park. There was no space for people like this, my tribe, to consume political news. Initially my focus was to build something for Black people. But then a friend introduced me to Robert Raben, a politico who happens to be Jewish and prides himself on running a majority-minority public relations firm in DC. He too wanted to create something like I had in mind, focusing on all communities of color. What a great idea. We connected and entered a contract to build *The Beat DC*, a daily news platform at the intersection of politics, policy, and people of color.

Daily newsletters have become a popular, reliable way to get information in people's in-boxes. Most major news outlets offer a daily newsletter, and there are platforms centered exclusively on the newsletter itself. For those who followed politics, *Politico Playbook* was a must-read in the morning. However, for those of us of color, the morning read often prompted two conversations: what was in *Playbook* and what was not. The same was true for *Axios*, another newsletter platform. Both seemed to frequently, and unintentionally, overlook topics relevant to communities of color. Then there was *TheSkimm*, which also seemed to speak to a mostly white audience. And all of these outlets were frequently sprinkled with dated culturally appropriated epithets, like the "Who Run the World" headers, with content that resonated with venture capitalists and mostly white consumers.

I was eager to create something authentic. Of course there would be the high-brow discourse about policy or the latest white paper from a think tank. But we'd also include a line or two about Colin Kaepernick and the NFL, comment on the latest episode of HBO's *Insecure*, or randomly drop a favorite line from *Black Panther*. Whether you were a political novice or a connoisseur, you could still enjoy the read and, most important, learn from it. It turns out that what was happening in politics through the lens of Black, Latino, Native American, and Asian American Pacific Islander communities was of great interest to many readers. By the end of 2017, after we had been up and running for a little over a year, tens of thousands of people all across the country were reading our content. I started getting notes from law professors in San Francisco about how much they enjoyed the newsletter. Entertainment executives in Los Angeles started resharing our content. Wall Street financiers began following our platform.

There was work to do. A lot of my friends had children, and when their kids were young, we lost touch. They would tell me there was literally no time to keep in touch—they were sleep deprived, irritable, couldn't keep up with texts and emails, and the child took every ounce of their energy. Well, I don't have children, but *The Beat* was my baby. And it was sucking away every minute of spare time I had. My days started at 4 a.m. and ended at midnight. I had a small team of two people, or sometimes one person. I was inundated with emails from hundreds of people across the country who all wanted their content in *The Beat*. I had to make sure our content was equitable and inclusive of all communities of color. Did we have too much Latinx content and not enough AAPI (Asian American Pacific Islander) stories? Were there enough photos of Native American politicos? Was our content about breaking down legislation moving through Congress gender-equitable? Then there were the hours of research, calls to Capitol Hill for quotes, and reading the latest white papers from think tanks to pull out relevant data for people of color. Every single day. I. Was. Exhausted.

I was doing what outlets with dozens of staff and tens of millions of dollars were doing. Though our platform was growing at an exponential pace, every venture capitalist I sat across from was incredibly dismissive of us. One VC firm suggested that I should go back to being an employee and give up being an employer. He advised me to simply fold my outlet under *Axios* since they could use the diversity. No thanks. While white men came in with a concept and were getting funded, we actually had a successful product and were getting declined. In 2017, venture capital investment reached $84.24 billion, a height not seen since the dot-com bubble of the early 2000s. But little of that money went to start-ups run by people of color, according to an analysis released in 2018. The study by RateMyInvestor and Diversity VC, which analyzed all publicly available venture-backed deals over a five-year period, found that the majority of funds went to college-educated white male founders. Researchers looked at the 135 largest US firms by the number of companies they invested in. Out of the 9,874 founders included, only 9 percent of startup founders were female; 17.7 percent identified as Asian American; 2.4 percent as Middle Eastern; 1.8 percent as Latin American; and just 1 percent identified as Black. On the funding side, studies have found that white men make up nearly 90 percent of venture capitalists. Hence, they're prone to fund companies run by similar people looking to address problems these funders find "relatable." Just 1 percent of VCs are Latin American, and only 3 percent are Black. This is how the infamous white man Billy McFarland could repeatedly get other white folks to write him checks for millions of dollars to fund a scam in the form of the Fyre Festival. Watch whiteness work. In 2016, the Center for Global Policy Solutions said discrimination and bias in favor of companies run by white men caused the United States to lose out on over 1.1 million minority-owned businesses and forfeit more than 9 million potential job opportunities.

This crowd was definitely not interested in funding people who looked like me. But despite the funding failures, television came calling. I had fallen onto the radar of cable news bookers. Just when I thought I was out . . . they pulled me back in again. Before long, they invited me to join the punditry class. This was going to be a new experience in a field I had long navigated. I was once again returning to cable news.

From the Control Room to the Greenroom

Where I had once been a flat fixture behind the scenes ripping scripts to eventually spoon-feeding reporters and anchors verbiage on cable news, I was now about to become a rounded face in front of the camera. I had essentially grown up in this space, so I knew it well. I had previously written scripts for countless anchors and reporters. I often drafted talking points for other people who appeared on television. I was a constant critic of some of what I'd seen displayed across my screen. I had to hype myself up and convince myself that going in front of the camera would be okay. But honestly, I did not think I would fit in as an on-air political analyst. There was no one like me in the space, and I wanted to avoid the expectation to be like my sister friends who had already made a name for themselves on the cable news circuit. I struggled with how I should present. How could I be expected to sit next to people whose research I used to pull and whose coffee I used to grab and now talk as their peer? After unsolicited advice from well-intentioned peers on how to present, I had to make a game-time decision. I was simply going to be myself.

I didn't want there to be any difference between the on-camera me and off-camera me: a Black face with an authentic Black voice capable of discussing issues of global importance, domestic policy, the media landscape, Capitol Hill and the current White House administration, campaigns and elections, civil rights and social justice, and more. As a Black woman, often I have felt the need to defend and define our humanity. I was happy to assume that role on air, but I also wanted to stretch beyond that in this medium. So frequently it can feel like our perspectives are obligatory must-haves to avoid the criticism of lacking diversity, only to be met with a collective eye roll from some who publicly urge Black people to move on from issues of race already. I was prepared for that. I was here to speak truthfully, and I was comfortable if that made others uncomfortable. It was not worth sacrificing my honest testimony to spare the feelings of those around me. I would not shuck and jive, nor would I downplay my own unique life experiences. I would not roll my neck and snap my fingers, aiming to go viral. I would have the simple goal of informing audiences on historical or present-day policy and shaping that information with my informed opinion. I had earned this seat, and welcome or not, I felt right at home.

My first time appearing on cable news was at Fox, where I felt less at home. I had been invited to join Harris Faulkner, one of their few Black hosts, for her afternoon show *Outnumbered Overtime*. Harris was always excited to have me on, and we always chatted pleasantly afterward. However, I was rarely excited to appear across from some of the fringe right-wing guests that populated the air. The last time I had appeared on the network, show producers told me we would be discussing the 2018 midterms. However, once I was seated and mic'd, Harris began reading copy that was all about North Korea. Huh? Had topics changed and no one told me? Yes, that's exactly what happened. Somehow the other guest, the conservative, knew the topic was North Korea, but no one had told me. This was not Harris's doing—she wants good TV with a healthy debate. If I sat there with nothing to say, it would not bode well for her ratings. So I suppose this was a tactic by someone else to make me look foolish. When will they learn? I took a deep breath at the cheap trick and was fully prepared to discuss North Korea. As I mentioned, I had been reading global news wires for twenty years and was currently writing about Trump's hypocrisy with the country's dictator, Kim Jong Un. You want to talk North Korea? Let's do this. As the other guest began spouting how Trump had won on North Korea, I quickly calculated in my head how many times Trump had lied about US–North Korea relations. When Harris tossed to me, I pointed out these lies and stated that Trump has a tough learning curve on the global stage. "Don't you think it was premature for Trump to say that North Korea was no longer an issue when, not long after that, people produced maps where Kim Jong Un is still building nuclear sites?" The response from my fellow panelists? "I don't want to get too deep in the weeds here but . . . blah blah blah." Eye roll. Okay. Bye. Farewell, Fox. You tried it.

I was excited when I was invited to return to CNN as an on-air analyst. I had the perfect plan. I was going to show them all that despite their treatment, I had survived. And all the things they knew twenty years ago—I know all of that and more now. So *ha*! I was going to spend more money than I could afford and wear some fancy outfit cloaked in an obnoxiously fashionable cape topped off with some oversize black shades (very Dominique Deveraux, with the added bonus of them not seeing my face until after I got my makeup done). I was going to take the long route out of the makeup room just so I could dust past all the people who had been awful to me and say, in my Joe Pesci *Goodfellas* voice, "Maybe you didn't hear . . . I don't shine shoes no more" and swing my cape as I walked away. I had secretly hoped that a booker from any other network would call while I was in makeup so I could loudly say, "I'm so sorry, doll. I am booked every night this week. I simply can't do that segment discussing the economic indicators of a recession." Yeah, ask me about Woodrow Wilson's Federal Reserve now, pricks. But damn! I never got the chance. Returning to CNN as an on-air analyst was *not* like walking twenty years in the past. The entire building had changed, so there was no conspicuously walking past people's offices to simultaneously flex and shade. Some people were still there, but many had left. Anchors whose research I used to pull barely remembered me. And there was a whole new crop of young people running around as guest greeters, learning the business just like I had. So I decided that I should exhaust some of that energy getting to know them instead of going on my petty walk of shade. Forward, Tiffany. Forward.

Just as most Americans don't have a full grasp of the functions of government, they also don't know much about cable news. Many have no idea how the behind-the-scenes process works. In the cable news space, you have hosts/anchors, reporters, contributors, and guests. The people who are introduced as contributors are people who have day jobs but get paid for appearing on the network. Those are the select few with deals contractually forbidding them to appear on any other network. Most talking heads want a contract. It's a matter of having another revenue stream but also having the distinction of being described as a "contributor." But deciding who should get a paid contract is a completely arbitrary process. To become a paid contributor, you must have a cheerleading squad. The hosts must like you, the bookers must promote you, the executives must be comfortable with you, and so on. Like newsrooms, most talent departments are run by white people, which, let's face it, can sometimes impact who is offered a contract and who is not. It's an interesting scenario to have more than 100,000 viewers cheer you on—but be overlooked and uncompensated because five to seven decision makers don't find you as appealing.

Makeupgate

Another space that is not inclusive for Black women is hair and makeup. Ask any Black woman who appears on air regularly, and I can guarantee you they have a nightmare story about a makeup artist making them look like a clown or a hair stylist having no clue how to handle their hair. This may seem like vanity, but it's not. People don't hear you at first; they *see* you. While appearing on TV, I could offer an insightful take on the Obama presidency awakening white rage. I could break down the difference between voter fraud and voter suppression. Or I could offer my thoughts and predictions on how the next three presidential elections will go. But if I did this with a foundation two shades lighter or darker than my skin tone, or an eyelash falling off, or a 1980s Farrah Fawcett hairdo, most people would not be able to focus on the substance of what I have to say, which matters to me.

Because the default with makeup artists is knowing how to glam white women, some Black women must do their own makeup. When Tamron Hall hosted NBC's *The Today Show*, she reportedly rarely let the on-staff artists do her hair *or* makeup. I am no exception. After having one too many unpleasant experiences (and a few draggings from Black Twitter), I began inquiring about having more people experienced with doing Black faces on the team. It's important to be clear here: in no way does this mean that only Black women can do Black makeup. Some of the best face beats I've had have been at the hands of Asian

American, Latinx, white, and Black makeup artists. It's not the ethnicity of the artist; it's their ability to adapt their skills to the ethnicity of their subject.

One time, I was set to appear on air to discuss the presidential election. Like many other Black female commentators who appear on this network, I began texting and calling the makeup artists who are skilled with doing Black makeup. None of them were working that day. I showed up to see someone I had previously only seen do hair. We looked at each other with the same level of consternation. I sat down in his chair, and thirty minutes later, he was done. I looked in the mirror. My right lash was only halfway on. My left lash was poking me in the eye. I was about to attempt to do something myself, but the producer was ready to take me to the studio. Was I really about to go on national TV?

I wanted my segment to be over as quickly as possible. I left a note on set. "Please hire artists who know how to apply a lash and can do Black makeup. Thanks." I had previously spoken to the woman who runs the makeup department. "Maybe this note will find its way to the right person," I thought. It did. The woman who oversees booking for the network's DC bureau gave me a ring. Before allowing me to speak, she said, "I saw the segment, and I thought you looked fine." Could this be because maybe she doesn't know what a Black woman should look like on TV? I told her I disagreed and informed her that it's a consistent problem. She said that she hadn't heard anyone other than me complain. Hmmm, that was strange. Everyone I knew who appeared on the network had complained to each other. I asked, "Have you ever bothered to ask any of the Black women who appear on your network what their experience has been? Have you ever asked the few Black talent that the network employs what their experience with makeup has been?" She clearly had not bothered to ask anyone who looked like me. She began interrogating me about who said what about their makeup and wanted to know who else I had spoken with. I was trying to answer all of her fifty-eleven questions. She went into an explanation that makeup artists there are not going to always have time to contour or pick the color you want perfectly. Another hmmm moment. They do for white women. Why not me? By the end of our conversation, her solution was that they should cancel all of my appearances on the network out of DC, since I was so uncomfortable, and that she would look into the matter. I took this as a threat and not an effort to address my concerns.

That obviously was not going to work. I told her I was slated to appear on the network that weekend. But I could let them know that, due to the makeup situation, your suggestion is to pull me from air. "No, do the show. I will watch to see how you look." I had to explain to her that I didn't anticipate any problems because the show's host, who happened to be a Black woman, had her own makeup team based on her past nightmare experiences. Perhaps, had she ever bothered to ask why this particular host requested a special team, she would have known the issue was ongoing. It was disconcerting. After appearing on the network multiple times a week, for free I might add, it was as if I should be grateful just to be there and dare not complain about how I look. It was a strange interaction, especially given that another woman who appears on the network regularly (a white woman who is a contributor) complained and received no pushback. In fact, they assigned her an entire hair and makeup team to herself, approved by her.

After the issue escalated, several people informed me that the woman I spoke with began warning other people within the company that I might be "a problem." Why? Because I tweet out things that are not good for the network. Puzzled, I asked the sources telling me this to ask her to produce the tweet. What was I saying that she allegedly thought was bad for the network? This was retribution. After I informed a few of the on-air hosts what happened, I got an email from her stating that she took our conversation seriously and had spoken with department managers.

A few days later, a network employee pulled me aside to explain to me that he knew about the note I left. He said that someone turned it in to an older white woman who works in the newsroom, who proceeded to make several disparaging comments about me, how I looked, and my audacity to complain. She is the one who then escalated *makeupgate* to the senior team that oversees who appears on the network. "She said some other things about you that I better not repeat or the network might be in some trouble," he said. Now what could this white man be implying that this woman said about me which was so awful the network might get in trouble? I won't bother to speculate, but I had a good idea. And I knew who the woman was: she oversees the entry-level desk assistants at the bureau. I wondered how her management and attitude had impacted young journalists, especially those of color, as they were navigating newsroom politics, just as I had some twenty-plus years prior. After months of no changes, like several of my colleagues, I resorted to hiring my own makeup artist when I appeared on the network—an expensive necessity exclusive to women of color.

Morning Joe

By 2019, I was completely independent. I had sold my house, gotten rid of my car, and consolidated debt so I could self-finance *The Beat*. Money was tight. And a Black woman running a media venture trying to secure financing from yet another industry dominated by white men had proven more difficult than I had imagined. So I was delighted to get a call from MSNBC's *Morning Joe*, which is considered one of the network's flagship programs. Though *AM Joy*, hosted by Joy-Ann Reid, pulls in numbers on par with the weekday morning show, the network sometimes appears to prioritize the one hosted by the husband-and-wife team, Joe Scarborough and Mika Brzezinski. I had been appearing regularly on *AM Joy* and sporadically on *Morning Joe*. In January 2019, a booking producer asked if I could appear on the show. "Of course," I replied enthusiastically. I was a frequent viewer of the show and often found that though they had a large list of daily guests and contributors, the analysis often needed more diversity, particularly African American women. I had secretly longed to be that diversity.

Are we talking about the latest tweet from the president? The perpetual breaking news from the White House? The new Congress that had just been sworn in? Whichever. I'm up for it. None of the above, the booker told me. That morning the writer Anand Giridharadas had been on the show opposite the conservative writer Noah Rothman, who had just released his new book, *Unjust: Social Justice and the Unmaking of America*, which attempted to draw parallels between Trump supporters and the social justice movements across the country making "victimhood" their common denominator. Giridharadas had been critical of *Morning Joe* on air, pointing out that a panel of all men had been chosen to discuss the principles Rothman argued in his book. Hence, the bookers wanted to make the next day's panel more equitable. Enter me.

The show's booker asked if I could make my way to New York City and join them on set. Absolutely, I replied. But in truth, I had no idea how I was going to "make my way" to NBC News headquarters. Every penny was accounted for. I know what you're thinking: How is this woman still broke after all these years? Didn't she just tell us about being broke twenty years ago? And she still can't afford a train ticket to New York and one night in a hotel? Damn! Yes, well—in that moment, I was asking myself the same question. It was already the afternoon, so I had to think fast. An Amtrak ticket to New York would cost at least $500 round-trip. Hotels were going for at least $300 a night. *Morning Joe* aired at 6 a.m. EST. Taking the first train out would be problematic, as I would have to run from the train to get on set with little time to prepare. I wondered how often the regulars on *Morning Joe* dealt with these issues. I assumed they had all built up enough money that dropping $800 on a quick trip to New York would not be a source of consternation. Or perhaps it would never occur to them to spend their own money to appear on the network. Perhaps they would just assume that because MSNBC was asking them to join, the network would incur the cost. I decided to do the same.

I was nervous about asking. I had waited a few hours to email the booker. It was the middle of the afternoon, and I had to be in New York in twelve hours. I decided to keep the email short and simple. "Hey, not sure how it works on *Morning Joe*. Do you guys do travel? Or should I plan to get my own hotel and train?" He replied back, looping in the travel department and asking them to handle booking everything. Whew! I got to New York just after 7 p.m. It was bitter cold. The air was so icy that inhaling hurt the inside of my chest. There were no cabs. The streets around Penn Station were oddly empty. I still had about five hours worth of writing to do for my own daily newsletter, *The Beat DC*. I was exhausted and overwhelmed, and felt ill prepared to do anything but sleep. It was just after eight when I got to the hotel. I tried to smack myself into conscious-ness, broke out my laptop, began writing my daily platform, and finished just after midnight. That's when I began thumbing through the pages the booker had sent me as an advance copy of Rothman's book.

At first, I just perused. I didn't think there would be any-thing so complex that I would have to memorize the passages. But as I read, I became increasingly frustrated at Rothman's account. Surely he wasn't suggesting that my idea of social jus-tice was a threat to American life? He was. Surely he wasn't suggesting that American political parties' ideologies weren't predicated on race until recently? He was. Surely he was not conflating the empty declarations at MAGA rallies to the years of oppression endured by Black people? That sure did appear to be his assertion. It was just after 9 p.m. on the West Coast, where my friend Albert Sanders was. He had worked in the Office of General Counsel under President Barack Obama and was a skilled debater. I rang him and barely let him say hello before I started reciting passages to him. "Albert, I want to go in on this dude tomorrow. Should I?" Albert responded the way I knew he would. "Hell yes. And don't let Joe interrupt you." I got about three hours of sleep, essentially a nap.

The next morning I felt ready. I had heard enough revisionist history my entire life that it was a reflex to course-correct people who wanted to paint the past in their colors. And lately, I'd had my share of people trying to "both-sides" anything having to do with Trump, his ideology, or his sycophants. Before heading to the set, I had done the mental equivalent of cracking my knuckles. I said hello to Noah in the greenroom and ribbed him for some of the things he detailed in his book. "Hey, Noah, how are your Twitter mentions?" I asked jokingly. "You can only imagine," he replied. I was on set before Noah to join Joe, Mika, and Willie Geist to talk about the news of the day. Everyone was mid-conversation by the time I made my way to the set, and they seemed unaware that I had been seated. "Hi Mika," I said. "Tiffany, thank you so much for being here. Welcome," she replied warmly. From that point on, I felt grateful that she was going out of her way to make me feel welcome. The first segment flew by. Now it was time to discuss Noah's book. For this discussion we were joined by Giridharadas, the conservative writer Reihan Salam, and conservative commentator Holly Harris.

Noah began philosophizing about "individuals" versus

"tribes." His goal seemed to be to water down claims of racial discrimination because "both sides"—white nationalists *and* people of color—make the argument that the meritocracy is out of their reach. He complained that antiracism and feminist philosophies and policies exert "negative pressure" that punishes —wait for it—white men! He went on to argue that affirmative action had created a new type of discrimination. "Philosophically, in this book, an action to positively discriminate, to rise up, sort of had a collateral damage effect of negatively discriminating," Rothman said. "That was an accident." Define negative discrimination, I asked. "Classes, whole tribes of individuals need to have a comeuppance, a due, sort of, reckoning, with their historical privileges that they know or may not know that they benefit from," Rothman said. "Likewise, to examine those who have been victimized by historical forces that they may either be aware of or not aware of. That is negative discrimination."

At this point, Joe asked whether he meant white men had been victimized by so-called negative discrimination. "It's a philosophy that views not individuals as individuals," Rothman said, "but as people who inhabit tribes, who inhabit a matrix of persecution in which they might not be aware of it, but this is a very powerful philosophy taught mostly on campuses, embraced by things like the women's movement, exemplified by things like intersectionality, which demonstrate, for individuals who may not be aware, that they have varying degrees of prejudice that they either suffer from or benefit from, and that is a sort of philosophy that I think is toxic because it makes you think of yourself and the people around you as not master of your own destiny."

I had to point out the numerous flaws in Rothman's argument. "Many people are not master of their own destiny," I retorted. "I think it's an inaccurate philosophy. With respect, I think your outlook on this is a bit myopic and based solely on your experience, and doesn't extend the intellectual debate with people who haven't had your history." I must say, Noah was at a bit of a disadvantage here. I'd heard these things for much of my life. I had observed the way that many white people, including the recently "woke" bunch, had existed in a bubble that gave little credence to my lived experience in this country. In that moment it felt like James Baldwin himself was whispering in my ear one of his many philosophical gems: "You give me a terrifying advantage. You never had to look at me. I had to look at you. I know more about you than you know about me." Hence, it was effortless to pick apart this ill-formed contention. "Certainly, there are economically disadvantaged white people in this country," I continued. "I think the reasons why they're socially and economically disadvantaged are very different from the reasons why some communities of color are disadvantaged and think it's really dangerous to look at the current state of this country without looking at it through the context of historical systems that put those people in that position. You call that victimization, but I call that reality for a lot of people."

It was a rare moment on the show. Noah and I went back and forth. He seemed to absorb the points I was making. I believe it was likely the first time he had heard someone articulate them. The other panelists quieted and let Noah and me have the floor. Joe, despite his gregarious nature, only interjected a few times, offering legitimate points as he attempted to enlighten Noah as a member of the same tribe. But the conversation mostly centered on the two of us, while the rest of the panel observed as though they were watching a professional tennis match unencumbered by an umpire accused of anti–Black woman sentiment (hey, Serena!). By the time we got off the air, we were the number three trending topic on Twitter.

Noah and I continued our conversation in the greenroom before departing. I got a note almost immediately after our appearance: "I hope you'll forgive the unsolicited email. Just wanted to reach out to thank you for the conversation we had today on the show around the issues in my book. I sincerely appreciate it. You were great, and your perspective was enlightening. The praise you're getting on social media is well deserved." It was signed by Noah. He also sent me a copy of his book. We agreed to meet for coffee when he was next in DC. It was a few weeks later that we met at a restaurant in the Jefferson Hotel just blocks from the White House. We shared a bit about our background and philosophies. Noah insisted that there's not as much daylight between us as I may think. Before we left, he said, "You should consider writing a book." He was the gazillionth person to express such a thought. "Maybe I will one day," I replied. I had many stories to tell and a lot to say.

3

Ring the Alarm!

Every generation of Black people has a crystalizing moment that punctuates this country's history with law enforcement violently claiming Black lives and subjecting our communities to subhuman treatment. For me, that moment was Rodney King. I was in seventh grade in Marietta, Georgia, when the King assault happened. White LAPD officers pulled King out of a car and beat him brutally while an amateur cameraman caught it all on videotape—this was before the era of ubiquitous smartphones. The four officers involved were indicted on charges of assault with a deadly weapon and excessive use of force by a police officer. But after a three-month trial, a predominantly white jury acquitted the officers, sparking the 1992 Los Angeles riots. Like many Black people, I had heard about these types of incidents from my family since I was old enough to comprehend words. But witnessing such violence on a screen was jarring to my thirteen-year-old spirit. Eventually, the story of King faded. But police violence against Black people did not. It would be hard to tell, however, because it was certainly not getting coverage.

This is one of the issues consistently ignored by white media, both nationally and locally. The only time it became a story was when the violence was so visual it was impossible to ignore. The brutality was not new. But smartphones with video recording capabilities were. Many people (not Black people) actually thought that perhaps there was an uptick in shootings leading up to the 2016 presidential election. They were wrong. There is a long American tradition of Black people being routinely terrorized by law enforcement—there was no uptick, just more coverage. And when this sea change occurred in the mainstream media, much of the coverage still lacked perspective and was reported on and discussed by largely white voices. Yes, there would be the obligatory Black analyst on air expected to provide historical context, unpack the specific violent incident, defend the unarmed victim, and educate the audience on this aspect of the Black American experience, all while surrounded by white people who were hearing this perspective for the very first time. Oh, and would you mind being entertaining while you unpack all of that? There's ratings to consider after all.

Sadly, by the time the media started paying attention, many Black people were already tuning out. You can take only so many segments of reporters getting slapped by storm winds when you're worried about getting slapped in the face with a nightstick by a cop. When former New York City mayor Michael Bloomberg ran in the 2020 election, it was the legacy of "stop-and-frisk" that helped to stifle his campaign. Though lawmakers in the Big Apple had given this policy a name, the practice of law enforcement officers routinely stopping and searching young Black and Latinx men—for no reason other than being young men of color—was common across the country. But it wasn't something that was widely covered by cable news outlets. The policy in New York City got a few mentions here and there on cable news outlets but hardly the exposé on the clearly racist policy that was warranted. Until, that is, Bloomberg ran for president. Black voters across the country, hearing of his policies and very familiar with the practice, delivered a decisive "no" to the wealthy white billionaire who had stepped up to try and take out the other racist white billionaire. When the arbiters of information reflect *our* American story, which disproportionately includes police brutality, the impact on democracy is immediate.

In the years leading up to the 2016 presidential election, only a few dozen out of more than 2,300 elected prosecutors nationwide were Black. The overwhelming portion of the elected officials responsible for many consequential decisions in the criminal justice system that have an inordinate impact on the lives of Black people—including whether to charge criminals, what sentences to seek, and whether to allow the accused to strike plea bargains—are white men, who make up a whopping 95 percent of the field. That means most charges brought against Black people, or not brought, are decided by people who do not look like us. And the systemic racism that permeates throughout prosecutors' offices across the country contributes to the racial disparities in the criminal justice system. So many innocent Black people habitually had their lives snatched by zealous prosecutors who saw their role as protecting white life, containing Black people, and stripping away their humanity. Not serving justice.

When the media decided, for the first time in recent history, that stories of police brutality were worth covering, Black people—having seen their blood spattered across screens—took a more active role in the democratic process. The civic engagement at the local level brought in sweeping changes, as municipalities across the country elected more Black prosecutors. Marilyn Mosby was sworn in as the Baltimore, Maryland, state's attorney in 2015, unseating a five-year incumbent and becoming, at thirty-five, the youngest prosecutor ever sworn into office in the nation's history. In 2016, Aramis Ayala was sworn in as a state attorney in Florida, as was Kim Foxx for Illinois. Kimberly Gardner took office in 2017 as the circuit attorney (chief prosecutor) for the City of St. Louis, Missouri. In Georgia, just outside Atlanta, Darius Pattillo was elected district attorney for Henry County. In rural Mississippi, two Black district attorneys were elected, as well as one in Caddo Parish, Louisiana, which is known as the nation's leading jurisdiction for death sentences. Democratic megadonor George Soros, through the racial justice group Color of Change and a network of Safety and Justice PACs, funneled nearly $5 million into electing more Black prosecutors and more than $16 million to elect prosecutors who want to reduce incarceration, crack down on police misconduct, and revamp a bail system they contend unfairly imprisons poor people before trial. The National Bar Association began urging Black prosecutors to run for office, and the National Black Prosecutors Association has offered to mentor candidates. Imagine if the white media landscape had made police brutality an ongoing story after the Rodney King assault. Our country's entire democratic landscape of locally elected prosecutors could potentially look different, which could have led to sweeping changes in our incredibly bigoted criminal justice system.

When the media ignores the Black experience and turns a blind eye to the consequences of white supremacy, the impact is not only democratic but economic as well. Take Ferguson, Missouri, where a majority of the population is Black—67 percent. The Department of Justice determined that the city had been operating a predatory system of government. This was only uncovered after the DOJ began investigating the deadly shooting of Michael Brown at the hands of a white police officer. Investigators discovered that police in Ferguson had been directed to generate revenue for the city's budget by using fines, fees, and various other tactics, targeting Black working-class people to extract money from them. In a city where the median household income is just over $41,000 a year, a single missed, late, or partial payment of a fine could potentially mean jail time. In one instance cited by the DOJ, a Ferguson resident was ticketed for parking her car illegally. That single infraction ended up costing her more than $1,000 and six days in jail. Payments from these types of encounters in Ferguson were pursued so aggressively that in 2013, outstanding fines and fees made up one-fifth of Ferguson's entire revenue base—an 80 percent increase over just two years prior. African Americans were funding a system that was oppressing them and bleeding them dry. These predatory tactics are not unusual, nor are they specific to any city. Com-

munities across the country have been practicing them for decades. But a University of Minnesota study found that there was
an uptick in 2007 during the Great Recession—which already
disproportionately impacted Black people's wealth—when tax
collections dropped due to the weak economy, and municipalities needed to find more sources of revenue to pay for ongoing
operations. "Cities, counties, and townships have turned bail,
civil asset forfeiture, and the like into major sources of revenue
for their municipal budgets," their research found.

This predatory practice of specifically targeting Black people
to fund the tax base is prevalent across many cities. People in
Black neighborhoods were certainly familiar with these methods. My own family knew them well. But these practices were
invisible to the master narrators and, as such, to American politics. It took a Black man getting shot by the police to force the
US media to pay attention to and cover what was happening in
the African American community. This then forced the DOJ
to investigate the shooting, which ended up uncovering a massive system of economic oppression that uniquely keeps Black
people impoverished. I know Trump's tweets are really, really
interesting—to the white folks who run newsrooms. But these
unethical practices are worthy of far more critical analysis than
they have received thus far.

TWO

How WE Got Here

4

The Trump Card

Anybody play chess? Yeah, me either (I really should learn, though). I'm more of a Spades girl. And in the card game of Spades, the trump card wins. In 2016, the Russians had the Trump card. But before they could obnoxiously smack it on the table, they had a clear directive: keep Black players out of the game entirely. I assume they figured Black voters in America would be easy pickings, given that we had been dealt such a bad hand and they had already secured a candidate who was a big joker. Black voters, despite a long, treacherous journey to the ballot box, had proven to be a powerful bloc that could potentially slap a wannabe king clear off the table. The Kremlin had to ensure that the surplus of Black Obama voters would renege in 2016.

Though Black voters primarily vote for Democrats, the media's constant use of "the Black vote" is a misnomer. There is no such thing as "the Black vote"—as if African Americans, who comprise 14 percent of the US population, more than 41 million people, could move as a homogenous group. We can't even agree on whether OJ did it or not, so we're definitely not in agreement on who to vote for. There are varying issues of importance, and sects of the Black community may prioritize policy differently. In fact, the most simplistic factors may drive Black voters in different ways. We're not talking police violence or education. Take something as simple as water. Everybody needs water, right? But water hits in different ways when Black folks are concerned. In the Lowcountry of South Carolina, Black farmers worry about how the government will respond to a drought. In California, Black voters in Long Beach might worry about a water shortage. And in Flint, Michigan, they have all the water they can use, but it's literally killing them. And that's just geographically.

African Americans have other issues driving our vote. Black folks in New York City, where the state taxes are considerable and the rent is too damn high, may vote their tax bracket. Working-class Black folk in coastal Louisiana may vote on environmental issues, given the catastrophic 2010 Deepwater Horizon oil spill. Black women have been enlisting in the US military at higher rates than any other demographic, and they represent nearly a third of all women in the armed forces; they may vote based on foreign policy. Given the diverse issues that activate and galvanize Black voters, a foreign adversary who's really trying to make a bid for the presidential election has got to focus on the one thing that's all but guaranteed to depress African Americans: racism. And because the media didn't cover state-sponsored violence against Black people with concern,

nuance, and consistency, the Kremlin picked that issue to infiltrate our democracy.

So in 2016, the Kremlin's Internet Research Agency (IRA)
punctuated this issue with a digital side-eye toward political
institutions while cultivating seemingly authentic narratives
of Black pride and bingo. A foreign adversary thus birthed a
significant number of African American followers in the absence of reliable media outlets to turn to for proper context and
analysis of what was happening. Russian operatives purchased
3,500 Facebook and Instagram ads to influence the presidential
election by exploiting the toxic race relations in America. Their
online army reached at least 146 million people, and Black folks
were the single group targeted most heavily—it wasn't a close
margin.

The media's framing on this is, however, incorrect. The issue
that attracted the Russians was not simply Black people; it was
white people and how they have historically discounted and brutalized Black people. That was the "in" for the IRA. According
to the *Guardian*'s the Counted Project, 573 Black people were
killed by police in 2015 and 2016 alone. It's impossible to tally
the exact number, since federal data of lethal police shootings are
notoriously incomplete. But a disproportionate number of Black
deaths at the hands of police officers emboldened by bigotry
ensured our television screens were bedecked with image after
image of dead Black bodies. This tragedy porn gave the cable
news outlets a perfect backdrop to the white outrage the Barack
Obama administration incited. And it gave the IRA the perfect
common denominator among most, if not all, Black voters: mis-
trust in law enforcement.

It is lazy, uninformed thinking and simplistic analysis to say Black voters abstained from participating in the 2016 election without offering some perspective. Consider, if you will, this split screen: On one side, a breaking news banner with a video montage of unarmed Black people being unfairly targeted and killed. On the other, white supremacists welling and swelling the red tide that aimed to "make America great again"—as if it had ever been great for Black people. This unfortunate collusion illustrated our timelines, television screens, and in-boxes. You may think you know well all the deaths that were happening. But I assure you, there were so many it's impossible to recall them all without a printed reference. This is a painful enumeration of those who needlessly lost their lives, like so many before them, simply for being Black—something the media could never bring itself to say. The first death that gut-punched Black people occurred just after Obama's reelection. It incited fury. And fear. Many Black parents hugged their children a little tighter and a little longer while mourning the appalling death of Trayvon Martin.

The mainstream media initially showed two pictures of Trayvon: one in a hoodie and one of him wearing fake gold teeth—both apparently menacing to white people. They referenced school suspensions and possible marijuana use. Much of the reporting presented the killing as a "tough call" type of situation and not the cold-blooded murder of a child. Trayvon was treated as an adult in the media, not as a teenager in high school. What the media never told you about the Florida teenager is that he dreamed of becoming a pilot or an aircraft mechanic. He had studied at George T. Baker Aviation Technical College in Miami during his freshman year of high school after being inspired by his uncle Ronald Fulton, who had a brief career in aviation. His friends and family described him as easygoing. But none of that seemed to matter. On February 26, 2012, Martin was gunned down by a wannabe police officer, twenty-eight-year-old George Zimmerman. The adult aggressor, who was armed with a 9mm pistol (which he later sold at auction for $250,000), had followed the teenager, suspecting him to be a burglar—though Martin was in the neighborhood where his father lived. Despite being told by a 911 operator that he was not needed to follow the teenager, Zimmerman confronted Martin and shot him. He claimed self-defense. Nearly all of Black America cried bullshit. Imagine a Black adult male pursuing and killing a white teenager. Would the treatment or media coverage be the same? Protests sprouted up around the country. America was in a state of unrest. And people wanted to hear from the first elected Black president—the one for whom we showed up in droves.

The first time President Barack Obama addressed the killing was in March 2012, when a reporter asked him to comment on the case. It was more than a month after Trayvon's murder. "Obviously, this is a tragedy. I can only imagine what these parents are going through," Obama said in the White House Rose Garden. "When I think about this boy, I think about my own kids." Perfectly humanizing comments for the president of the United States to make. But for bigots, his next comments were offensive. "If I had a son, he'd look like Trayvon," Obama said, pausing for a moment before he continued speaking about Trayvon's parents. "I think they are right to expect that all of us as Americans are going to take this with the seriousness it deserves, and we are going to get to the bottom of exactly what happened."

Obama had dared to acknowledge his Blackness. So for all the voters who cast ballots for Obama in spite of his being Black, this felt like a betrayal. It wasn't just the crazy fringe wing of super conservatives who found Obama's remarks offensive. It wasn't just the conspiracy theorists who post nutty comments on internet articles who were bothered. It was the closeted racists who never considered themselves white supremacists. After all, they voted for Obama. Ergo, it was okay to view certain Black people as deserving their fatal outcomes. Perhaps they tried to suppress the thought "The kid probably had it coming." They hadn't yet been emboldened to express those thoughts aloud. But soon, a new president would let them know that not only were their thoughts valid, but they were welcome in the new "Great America." In summer 2013, a jury acquitted Zimmerman of second-degree murder and manslaughter charges, prompting Obama to speak again.

"You know, when Trayvon Martin was first shot I said that this could have been my son. Another way of saying that is Trayvon Martin could have been me thirty-five years ago," Obama said. "There are very few African American men in this country who haven't had the experience of being followed when they were shopping in a department store. That includes me. There are very few African American men who haven't had the experience of walking across the street and hearing the locks click on the doors of cars. That happens to me—at least before I was a senator. There are very few African Americans who haven't had the experience of getting on an elevator and a woman clutching her purse nervously and holding her breath until she had a chance to get off. That happens often." He continued, "And I don't want to exaggerate this, but those sets of experiences inform how the African American community interprets what happened one night in Florida. And it's inescapable for people to bring those experiences to bear. The African American community is also knowledgeable that there is a history of racial disparities in the application of our criminal laws—everything from the death penalty to enforcement of our drug laws. And that ends up having an impact in terms of how people interpret the case." Some wanted Obama to be more direct and say the only reason this child is dead today is because he committed the crime of being born Black in a country that did not value his life. But many white people felt abandoned and betrayed by the Black guy they had been able to stomach. The special magical negro who came to the world through the womb of a white woman had just disrupted their comfortable afternoon, where they enjoyed an existence oblivious to these types of inequities. After all, the country was supposed to be "post-racial." The seed of quiet white rage (which is always there) got watered that day and began to sprout once more. Meanwhile, Zimmerman continued to torment the Martin family by suing them in 2019 for $100 million in damages.

After this, the shooting stories seemed endless. The next child murder came just months after Trayvon's tragic death. This one was also in Florida and also involved an adult man and a seventeen-year-old high school student. It was the day after Thanksgiving, Black Friday, when Jordan Davis, a passenger in a car with friends, pulled into a Jacksonville gas station. The music playing from their car bothered a forty-five-year-old white man, Michael Dunn. He asked the boys to turn it down. They complied but not enough for Dunn, who grabbed the 9mm pistol in his glove compartment and fired ten shots at the children, killing Davis. Although he fired the final shots as the vehicle sped away, Dunn claimed at trial that he was acting in self-defense. He claimed Davis had brandished a gun and moved to attack. Police found no weapon, and witnesses said they never saw one. Yet countless times in the media the question was posed by reporters, anchors, or in news banners: "Did he have a weapon?" Dunn was found guilty of first-degree murder. Davis's mother, Lucy McBath, now represents Georgia's sixth congressional district.

In July 2014, Eric Garner died in New York City after police officer Daniel Pantaleo put him in a chokehold while arresting him on suspicion of selling single cigarettes from packs without tax stamps—a petty crime that resulted in the death penalty. The incident was captured on a cell phone. Garner, a father of two, told the police that he was tired of being harassed and he was not selling cigarettes. After Pantaleo removed his arm from Garner's neck, he pushed the side of Garner's face into the ground while four officers moved to restrain Garner, who repeated, "I can't breathe" eleven times. He died on the concrete. None of the officers were convicted of any crimes, despite the medical examiner ruling Garner's death a homicide. Though almost everyone found the outcome ludicrous, too few were honest enough to mention race. The killing of an unarmed Black man went unpunished because of the legal protection of his killer, a white police officer. Yet the media was hesitant to declare what was so obvious—that race played a significant role in Garner's death and the ensuing response. He was killed because he was Black.

The next month, John Crawford went to a Walmart in Beavercreek, Ohio, to buy marshmallows, graham crackers, and chocolate to make s'mores at a family cookout. While shopping, the twenty-two-year-old father of two picked up an unpackaged BB pellet air rifle in the store's sporting goods section. A customer called 911, claiming that John was pointing the gun at people walking by. Released surveillance footage shows that Crawford never pointed the gun at anyone, and the customer later recanted. For carrying a BB gun in the store that sold the gun, in an open carry state, police officer Sean Williams shot Crawford on-site. He died at the hospital soon afterward. Several witness statements negated the notion that Crawford was perceived as a threat by other shoppers. A grand jury decided not to indict the officers. When this shooting first happened, it didn't get a lot of national news coverage. It was as if the mainstream news media only had so much room for dead unarmed Black people, and they had enough to fill the airwaves.

Later that same month, eighteen-year-old Michael Brown was fatally shot by police officer Darren Wilson in Ferguson, Missouri. As I referenced earlier, this shooting helped uncover massive economic oppression in Ferguson. Wilson said that after Brown attacked him in his police vehicle, an altercation ensued for control of Wilson's gun, until it discharged. Something to note here: not too many sane Black men attack police. Whenever officers say this, it begs belief. A witness who was with Brown stated that Wilson initiated a confrontation by grabbing Brown by the neck through his car window, threatening him and then shooting at him. The shooting was deemed justified by both a grand jury and a Department of Justice investigation. The killing kicked off weeks of unrest that sparked standoffs between officers and protesters. Viral videos and photos showed police operating military vehicles to threaten crowds and firing tear gas and rubber bullets at lines of protesters and reporters. One reporter referenced Brown as "no angel." Again, the media's treatment of Black victims proved to be harsher than it is of white people actually accused of crimes. Some reporters from national outlets started paying more attention to the Twitter conversations about Ferguson, largely made up of Black folks who were bypassing the outlets who ignored them. Far too many reporters were relying primarily on elite, established sources that were painfully out of touch.

By November that same year, twelve-year-old Tamir Rice was playing with a toy gun in his Cleveland, Ohio, neighborhood. A nearby witness called 911 to report that someone was brandishing a gun in a park. The caller told the dispatcher the person was "probably a juvenile" and the gun was "probably fake." But upon arrival, twenty-six-year-old police officer Timothy Loehmann—who was still a trainee at the time—demanded the child show his hands. Within two seconds, he had shot the boy. Rice died the next day from a gunshot wound to the torso. It was later revealed that Loehmann, in his previous job as a police officer in the Cleveland suburb of Independence, had been deemed emotionally unstable and unfit for duty. A Cuyahoga County grand jury declined to bring criminal charges against him. The reaction from bigots was so visceral, the local paper, the *Cleveland Plain Dealer*, chose to permanently shut down their comments section to avoid the racist diatribes that peppered their site rather take on the responsibility of holding racist people accountable by letting others see their unfiltered thoughts. Their intentions may have been good, but it's still whitewashing the vile and disgusting outlook shared by too many people. And however good their intentions, some of their reporting was terrible. One article, titled "Tamir Rice's Father Has a History of Domestic Violence," was published four days after the shooting. Later the outlet ran an article headlined, "Father of Cleveland Cop Who Shot Tamir Rice Says His Son Had No Choice." Both suggested that somehow the twelve-year-old victim had played a role in his own death, but the officer who shot him was doing his duty.

In April 2015, Baltimore police officers arrested twenty-five-year-old Freddie Gray for possessing what the police alleged was an illegal knife under the city's law. They then shackled him, violated department policy by failing to strap him in a seatbelt, and took him on a "rough ride." They were trying to hurt him. The driver of the van intentionally took sharp turns, allowing Gray's body to be thrown around without the ability to hold himself steady, according to some. He died from injuries sustained to his neck and spine while in transport in a police vehicle. Spontaneous protests started after Gray's funeral, kicking off several violent incidents. At least 250 people were arrested, and thousands of police and Maryland National Guard troops were deployed. Yet much of the national news coverage around this tragic incident focused on two women: Baltimore state's attorney Marilyn Mosby, whose speech announcing charges against the officers went viral; and Toya Graham, a Black mother whose beating of her young son protesting was caught on camera. Never mind the collective celebration of many in white America who cheered this on, thinking the only way to raise a Black child is to beat him into submission. It was the fear and ferociousness of a Black mother who was desperate to keep her child alive, but they couldn't understand this through their lens. This is another example of how mostly white males who make the editorial decisions pick and choose the narratives. They frequently disregard authentic Black experience and focus on the wrong thing entirely.

Three months later, twenty-eight-year-old Sandra Bland was pulled over by Officer Brian Encinia in Prairie View, Texas, for a failure to signal when changing lanes. She said she was changing lanes because Encinia was tailing her so closely she thought he was trying to get by. Bland, who had been recording a series of videos about police mistreatment of Black people, had the foresight to record the incident on her mobile phone. The footage was not released until nearly four years later, but Encinia's dashcam captured much of the encounter. The trooper told Bland to "get off the phone," and Bland responded, "I'm not on the phone. I have a right to record. This is my property." Encinia, getting angry, ordered her, "Put your phone down. Put your phone down right now." He was contorting in anger as he pulled out a stun gun and shouted at her to get out of the car. "I'm going to light you up!" he yelled. He removed her from her car, forcibly placed her on the ground, and arrested her. Three days later, she was found dead, hanging in her cell. No one knows exactly how Bland died; her death was ruled a suicide. But we do know there was no reason for her to be arrested.

Some reports described the arresting officer as being courteous and Bland being "terse."

The following July in 2016, thirty-two-year-old Philando Castile was pulled over while driving in Falcon Heights, Minnesota, because a police officer, Jeronimo Yanez, said that Castile matched the description of a robbery suspect from several days earlier. Despite Castile doing everything right—he told the officer he had a firearm (which he was licensed to carry)—none of it made a difference. Castile was shot seven times just thirty-eight seconds after Yanez first approached his window. His girlfriend, Diamond Reynolds, was in the car with him and had the emotional wherewithal to capture the devastating moment on Facebook live. As Castile died, Reynolds's four-year-old daughter could be heard in the backseat trying to calm her crying mother, saying, "It's okay. I'm right here with you." Yanez was charged with manslaughter and reckless discharge of a firearm. He was acquitted of all charges. When Castile's mother appeared on CNN, she was asked if she had any thoughts about the shooting of a policeman in Dallas—as if this grieving mother had anything to do with that.

Just two months before Election Day—forty-year-old Terence Crutcher, a father of four, was shot and killed by police officer Betty Jo Shelby in Tulsa, Oklahoma. Footage from multiple police cameras shows Crutcher with his hands raised, walking toward his abandoned SUV in the middle of the road. Shelby and another officer approach Crutcher, who continues to walk toward his car. Circling above the scene, a police officer in the helicopter can be heard saying Crutcher looks like a "bad dude." Shelby shoots Crutcher when he remains unresponsive to her commands to stop walking toward the vehicle. She says she fired because she feared for her life, though Crutcher's hands were raised the entire time. Shelby was acquitted of first-degree manslaughter. She appeared on CBS's *60 Minutes* and told correspondent Bill Whitaker, "I have sorrow that this happened, that this man lost his life, but he caused the situation to occur. So in the end, he caused his own [death]." She was later tapped by the city to teach a course on how to "survive such events"—legally, emotionally, and physically.

The anger in much of the Black community was palpable. The ignorance and insensitivity of those in the mainstream press was grossly unrelatable. This was the perfect time for Russia to infiltrate media and disrupt our political system by tapping into deeper racial divides. They had a useful idiot in position to help drive the divisions and a host of unknowing coconspirators primed to help their cause. According to a New Knowledge report, the IRA campaign was designed to not only exploit societal fractures but erode trust in media entities and the information environment. By keeping Black faces and Black voices out of editorial decision making, the press helped the Kremlin pursue their ultimate goal: depressing Black voter turnout. They were well studied and well prepared. They had gathered stolen identities of real Americans, and had a formidable playbook of what resonates on social media when it comes to riling up Americans talking about politics.

They operated several social media accounts that repeated messages of police violence and voter fraud. Their internet bots spread fake news and pro–Donald Trump propaganda on Face-book, Twitter, and other social media platforms. According to federal indictments, the IRA controlled an Instagram account called "Woke Blacks." One of the account's posts in October 2016 read, "A particular hype and hatred for Trump is mislead-ing the people and forcing Blacks to vote Killary. We cannot resort to the lesser of two devils. Then we'd surely be better off without voting AT ALL."

Two members of the agency traveled to the United States for the sole purpose of gathering intelligence and embarked on a fact-finding tour in nine states, according to investigators. If, after spending time in America, the takeaway from spies is pretty much, "Yeah, they treat Black people poorly there. That should *definitely* be our angle," it begs the question why it's so hard for actual Americans to recognize the same thing. It's not that Black people were any more susceptible to Russian influ-ence than, say, Donald Trump for instance. It was our history with this country that left us open to voices acknowledging our struggle, loss of life, economic disparities, and suffering. So eager to build community around our collective pain, some were open to voices that were later proven to have nefarious intentions. The two spies were advised to aim for battleground states—states with swing voters who could declare for Repub-lican or Democratic candidates in presidential election years—such as Michigan, Pennsylvania, and Wisconsin. The IRA had a monthly budget of more than $1.25 million by September 2016 to carry out its influence campaign, according to charges filed by the Department of Justice.

We'll never know the precise impact these Russian troll farms had in shaping US elections. But Black voter turnout in a presidential election declined in the United States in 2016 for the first time in twenty years. A 2018 analysis of voter records found that about 4.4 million people who voted for Barack Obama in 2012 failed to vote in the 2016 election. Of that group, 36 percent were Black. That adds up to about 1.6 million people.

The way elections are supposed to work is that party candidates make a case to voters. Much of their case is filtered by the media. And when the media only covered white people's issues and not ours, that gap made way for the Russians to move in at a time when we had a powder keg of state-sponsored violence against Black people that had never been properly covered. It doesn't mean any one entity was singularly responsible for depressing Black voters. But if the only time we saw outrage expressed that Sandra Bland or Tamir Rice had been murdered was in a meme, then, hell yes, that would attract our attention.

I am also not discounting the fact that more than 62 million mostly white Americans cast a ballot for Donald Trump knowing full well who he was. They saw the racism, misogyny, and xenophobia, and their response was, "Sounds good to me. Let's make America great again." White supremacy has always been America's greatest weakness. Where the United States once pretended to be a champion of the poor, a nation drawing strength from its promise of dream fulfillment for everyone, it's now unmasked, revealing its white and "Christian"-sponsored bigotry, both indifferent to and dependent on African Americans. This fire was not started by the Kremlin. They were simply the gas lighters. But they had long spotted the chink in American armor. It was one they had pierced before, under eerily similar circumstances.

5

The Red Summer

Jordan Peele's 2017 classic *Get Out* brilliantly amplified America's ingrained racism via the horror film genre. One of the killers in the film says, "I would have voted for Obama for a third term, if I could. Best president in my lifetime, hands down." That's often how some white people present themselves, despite holding some prejudiced thoughts. For those not in the sunken place, these people are pretty transparent. Black people laugh, we smile, we tolerate—but at the end of the day we're typically fully aware with whom we're dealing. And as bothersome as these people can be, this was not always how "polite" racist white folks engaged with Black people. Oh no, no, as the housekeeper Georgina says in *Get Out*. No. No no no no no. During the Red Summer of 1919, when numerous white mobs were attacking Black people all across the nation, African Americans were living under constant terror. The media, even then, ignored the sheer violence white people inflicted on their fellow countrymen. But one hundred years ago, there was a thriving and vibrant Black press. And they were as informative as they were lifesaving, sounding the alarm for Black people throughout the south: get out!

As knowledgeable as I thought I was on this particularly violent period in history, what I didn't know still floored me. As much as the lives of Black people remain at risk when it comes to police violence, it's nothing like being under constant threat from law enforcement and vigilantes. And also, why is it that for much of my life I knew so little? I never learned about this period in school. Looking back, most textbooks omitted any stories of oppression at the hands of white people. When they weren't peppered with outright errors, they were mostly filled with erroneous interpretations. Pedagogy lent itself to alternative narratives of docile Black civil rights leaders preaching nonviolence, and conveyed very little of the cruelty endured by the folks on the receiving end of all that violence. By attempting to erase this time period in history, the media and politicians alike left themselves vulnerable to the weaponry used by the Kremlin. When white oppressors were ending Black lives in violent sprees throughout America over a century ago, it was the first time Russia took notice of what white newspapers largely ignored. This wrinkle in time denotes the striking similarities glued together by one hundred years, thousands of lost lives, and a segregated media landscape.

Prior to TV, there was a lot more decentralized media coverage. But because of advances in technology, globalization, and the growth of megacorporations, most of our news now is established by three twenty-four-hour cable networks and a slew of locally owned stations largely operated by the conservative conglomerate Sinclair Broadcasting. But during the Red Summer of 1919, there were hundreds of Black newspapers small and large throughout the country. Hence, it didn't matter if the white media was covering the violence Black folks experienced or not. The Black press was all over it. Now there are few left, and local Black news, which could once cover daily in-depth issues in Chicago, Baltimore, Cleveland, and St. Louis, are a shell of what they once were. The name "Red Summer" was coined by NAACP field secretary James Weldon Johnson, which the Black press reported. African Americans gave it a moniker, cementing it in our history whether the master narrators acknowledge it or not. This is much the same way that Black Lives Matter became a discussion point: in the midst of people dying, Black women dared to attach a battle cry to force the masses to take notice. As it has always been throughout history, it takes Black people dying in spectacular fashion to pierce the white narrative of what happens to us.

In the 1910s and 1920s, violence committed by whites was so abhorrent and detestable that civil rights commissions had to be set up in cities to pressure local and federal governments to address it. This is in part how police became the intermediary for what the white community wanted. It was okay to beat Black people to death, or to lynch them, or to shoot them—but you had to be wearing a badge while you did it. This also offers some explanation as to why there are some white people who deputize themselves as law enforcement when encountering Black people—also not a new phenomenon, despite this despicable behavior increasingly being caught on tape. From the

Yale college student calling the police on her fellow classmate for sleeping in the common room of a graduate student dorm to a white woman calling the cops on a Black man for his dog humping hers—in a dog park!—these particular white people assume that law enforcement is there strictly for them, and that when police show up they will instinctively be on their side. This is a deeply rooted sentiment of superiority over African Americans, which has been nurtured by an unforgiving criminal justice system that favors white people. For centuries, Black people lived in a perpetual war zone, where the police were never there for their protection but rather for their persecution. The year 1919 was no different. There may have been Black press to report on what Black people were enduring, but there was still a great white barrier stopping African Americans from participating in democracy. For the hundreds of thousands of Black military men who had shed blood to preserve the very thing they were denied, there was no reprieve from violence.

Losing Battles, Winning Wars

The year 1919 was blanketed in red; it wasn't just the summer that caused a scare. Black men who had just fought in a war— not for American exceptionalism, but for the vacant vow of equality—were returning home draped in the hollow promise of patriotism. But they would soon discover that conflict overseas was no match for the combat awaiting them at home.

It was just one year after the end of the First World War. Black people were driving music culture with ragtime sounds disseminated by brass bands—the entrée for what would eventually be widely consumed as jazz. There were no saloons in which to toast. Alcoholism, abuse, piety, and discrimination had brought on prohibition, so Black people who drank, particularly in the south, met up at the occasional juke joint. There was more foot traffic than cars on the road in 1919. There were only 6.7 million automobiles in the entire country that year (to put this in context, one hundred years later there would be 267 million cars on US roads). No cars meant a lot of walking, which is likely why women were not rocking high heels back then—nearly all wore lace-up or button-up two-tone black and brown ankle boots that would set you back anywhere from $5 to $14. But women did crown themselves in a bevy of fashionable hats and cloak themselves in loose-fitting petticoats, those who could afford them. This was the scene that welcomed home the roughly 380,000 African American soldiers who had served in World War I. But America would not greet these patriots who had just served their country as heroes. Sadly, they weren't even treated decently as soldiers.

In fact, some military historians have noted that Black men were only allowed to serve when War Department planners realized that the standing army of 126,000 men would not be enough to ensure victory against Germany. In 1917, Congress passed the Selective Service Act requiring all male citizens between the ages of twenty-one and thirty-one to register for the draft. White newspapers had been spreading rumors of Black soldiers assaulting white police since Reconstruction ended. Ergo, states across the south prohibited Black people from handling weapons—except when they needed their help saving the country. It's funny how gun laws in this country adjust depending on who possesses weaponry. Black men innocently thought that joining the army would garner them respect from their white counterparts. Perhaps then, they thought, they would finally be treated as equal citizens. Black periodicals like the *Chicago Defender* and the NAACP's *The Crisis*, which were largely focused on the rampant lynchings occurring in the south, discouraged Black people from joining the white man's war. But the intellectual W. E. B. Du Bois initially viewed Black people's participation in the First World War as potentially revolutionary. "I felt for a moment during the war that I could be without reservation a patriotic American," Du Bois reflected in his 1940 book *Dusk of Dawn*. "I am less sure now than then of the soundness of this war attitude." Du Bois, and the soldiers who believed as he did, was wrong. Even the process of signing up discounted Black lives and shuffled them to the frontlines, only to falsely brand them draft dodgers. The managing editor of *On Point*, an Army Historical Foundation publication, Jami L. Bryan, writes:

> It was fairly common for southern postal workers to deliberately withhold the registration cards of eligible black men and have them arrested for being draft dodgers. African American men who owned their own farms and had families were often drafted before single white employees of large planters. Although comprising just ten percent of the entire United States population, blacks supplied thirteen percent of inductees.

Black draftees were routinely humiliated and treated with aggressive animosity when they arrived for training. They were greeted by white men who refused to grant them any humanity whatsoever. The intolerant behavior was either overlooked or celebrated. Black soldiers found no unity among their country-men in the army—they remained rigidly segregated. Bryan notes that some were forced to sleep outside in tents instead of in warm barracks. They rarely received proper attire and could go months without a change of clothes at all. Some were made to eat outside like animals. They were given inferior medical care. And they were largely relegated to labor duties and had to confront slander from white soldiers and officers. Still they served. And they helped America win the war.

After these veterans labored in the army, weathered the abuse, and shed blood for democracy abroad, predictably, there was no democracy awaiting any of them at home. Upon their return, the men discovered that American white supremacy was harder to collapse than the German army itself. Many Black veterans were denied the benefits and disability pay they had been promised. And the mere sight of a Black man having the audacity to wear his uniform in public could incite violence from white aggressors who never wanted Black men to fight in the army. African Americans had returned from fighting only to fight once more.

Their heads may have been bloodied, but they were defi-antly unbowed. After shedding blood for the country that was actively oppressing them, they would be damned if they were going to continue to tolerate being treated as second-class cit-izens. Their service to the country had given them a newfound sense of patriotism. Despite every effort to make Black people feel unwelcome in a nation they had not only built but fought for, African Americans were declaring themselves equal. Black servicemen spoke out against racial discrimination, violence, and inequality.

As you can imagine, this audacity did not sit well with the white patriarchy that still ruled violently over the nation. And this patriarchy was aided by an all-white press corps pretty much encouraging the violence. Many of the white-owned newspapers in the nation's capital reported as truth fabricated instances of Black men assaulting white women. The *Washington Post* ran a front-page story advertising the location for white servicemen to meet and carry out further attacks on Black people in DC. As Black soldiers returned home, fanning throughout the country, many veterans were lynched. Many more survived brutal beatings and sadistic shootings. They had committed the crime of being Black patriots in the white man's stolen land.

The uptick in white violence toward Blacks at this time was not only due to a fear of losing control over African Americans; it was also the result of an increase in new immigrants coming to the country. Sound familiar? Anti-immigration racism and xenophobia are nothing new. White supremacy in America is like the iPhone that way: new versions keep coming out. Each new version does all the same things as the last, but it has a few new added features. There's always some bugs to tweak in the more modern operating system, but it eventually functions as designed. Which, incidentally, brings us to inhumane labor practices.

The American Rust Belt was experiencing severe labor shortages at this time. There was an influx of people arriving at the border in 1919, chasing the American dream, and many white Americans were not happy about the new arrivals. These immigrants were European, but they came from the "wrong" part of Europe: they were Slavs and Jewish people, who both faced widespread hostility. There were no Make America Great Again hats at that time, but if there were, the US Congress's new Republican chair of the House Committee on Immigration and Naturalization, Albert Johnson, would have most certainly sported one. He was a newspaper owner and an adamant supporter of the anti-immigrant movement in 1919. He wrote in one editorial that "the greatest menace to the Republic today is the open door it affords to the ignorant hordes from Eastern and Southern Europe." He introduced a plan that would allow immigrants to be admitted on the basis of country-by-country quotas that heavily favored Northern Europeans, while drastically reducing immigration from the less-favored European countries. The law was enacted in 1921 and would stand until 1965.

At this same time, northern manufacturers needed workers and found them in the first wave of the Great Migration, when hundreds of thousands of Black Americans fled the south and moved north. And despite the perception that crossing the Mason-Dixon Line opened a promised land for Black people, that was not at all the case in the northern states. African Americans still encountered segregation, racism, and inequality. Black workers were filling new positions in expanding industries, such as the railroads. They were also securing jobs once held exclusively by whites due to the 1919 steel strike, which handicapped one of the nation's largest industries and biggest employers in the country at the time. The strike took more than 365,000 workers off the job and onto the picket lines.

The labor dispute centered on white workers wanting better wages, job protections, and improved conditions. But the steel industry rebuffed their demands and refused to recognize unions, which Black people were not allowed to join. So between thirty thousand and forty thousand Black and Mexican American workers were brought in to work in the mills. Everyone *should have* been mad at the same people. But company officials, who were no friends to the Black community, knew the white workers needed someone to blame. So they stoked the racism and painted an inaccurate picture of how well fed and happy Black workers appeared now that they had "taken" jobs from white people. Years later, President Lyndon Johnson said, "If you can convince the lowest white man he's better than the best colored man, he won't notice you're picking his pocket. Hell, give him somebody to look down on, and he'll empty his pockets for you."

For the first time in history, ill-treated African Americans had the potential to exercise workers' power at the center of US industry. But they were denied labor rights, they were being attacked when they showed up for work, and they were desperately trying to build lives that white assailants attempted to destroy on a daily basis. The oppressive fist was tightening its grip around the neck of any Black person who dared to exist freely; so too were the clenched fists of Black people, who were ready to strike back. It was just a matter of time before things came to a head. And when they did, blood spilled. The killings, dangers, grievances, hardships, and anger stretched from the Midwestern Rust Belt and bled into the burgeoning north and back down to the blood-soiled south. It's impossible to recount all the violence. In fact, it's difficult to stomach the details. Yet it's a necessary reminder of what Black people in this country have endured and survived.

Our Blood Stripes the Flag Red, Our Tears Stain It Blue

For simply fighting in the First World War, declaring themselves equal citizens, and demanding to live free of violence and function in American democracy, Black Americans encountered a response from much of white America that amounted to "Hell no." The response was violent, cruel, and barbaric.

In December 1918, Charles Lewis began his last day as a private in the United States Army. As told in *Lynching in America: Targeting Black Veterans*, a report by Bryan Stevenson's Equal Justice Initiative, Lewis had accepted his honorable discharge and left one of the few military facilities that housed Black soldiers. He was leaving Ohio and heading home to Alabama. While he was waiting for the southbound train to leave, a deputy sheriff boarded the train car, looking for suspects in a robbery—an all too familiar excuse to stop most any Black person. He approached Lewis, demanding to inspect his baggage. Lewis declared that he had just been honorably discharged and had never committed a crime in his life. An argument broke out between the two, and Lewis was charged with assault and resisting arrest. He was taken to the county jail in Hickman, Kentucky. By midnight, a mob of as many as one hundred men gathered outside the jail. The masked men stormed the station, smashed the locks with a sledgehammer, pulled Lewis from his cell, and hanged him. His body was left for all to see. His lifeless frame was still draped in his United States Army uniform as he swung from a tree.

The first documented eruption of 1919, as detailed Cameron McWhirter's book *Red Summer: The Summer of 1919 and the Awakening of Black America*, would happen in Jenkins County, Georgia, about three hours south of Atlanta. The terror began on April 13, when a wealthy Black community leader named Joe Ruffin and his son, Louis, were en route to a festival at Carswell Grove Baptist Church, where the elder Ruffin was to serve as marshal. On the road, Joe saw a friend sitting in the backseat of a police car and stopped to offer to pay the friend's $400 bail by check. At this time, being in the back of a police car did not simply mean an arrest. It frequently meant a painful, violent, public death. Hence, Joe was not simply offering to pay his friend's bail; he was offering to buy his friend's life. The two white police officers refused Ruffin's check and demanded cash instead—something that, I'm sure the officers knew, was an impossible sum to get on a Sunday. Joe, desperate to save his friend, reportedly moved to collect him from the police car and attempt to save his life. The officer pistol-whipped Joe, and the gun discharged. His son Louis, thinking his father had been killed, swiftly shot the officer. Then all three of the Black men moved to defend themselves against the second officer, who pulled out a gun. They killed him as well, but not before the officer shot and killed the friend who had been detained.

In an essay for the Harvard Divinity Bulletin, McWhirter writes,

A white mob quickly formed and went on a rampage. The mob burned the church down, then killed two of Ruffin's sons—one of them a thirteen-year-old. Rioters threw the bodies in the flames, then spread out through the area, burning Black lodges, churches, and cars. They killed several other people; no one knows how many. The wounded Joe Ruffin was saved from the lynch mob only because a white county commissioner drove him at high speed to the nearest big city, Augusta, and put him in the county jail there.

Ruffin was charged with the murder of the two white officers and for months was threatened with lynching. No one was ever charged with the killings of his sons, the destruction of the church, or other crimes against African Americans throughout the county.

Louis and Joe Ruffin escaped with their lives, but not their livelihoods. They both died in exile. Guilty of murder, Louis fled from Georgia and went into hiding. His father died a free man in South Carolina but was impoverished by the legal fees that kept him out of prison and separated from his three remaining sons, his home, and their community. That was the first of many.

Across the way in Chicago, Eugene Williams and his friends were looking to cool off from a sweltering heat wave in the city. It was a Sunday and 96 degrees outside. The seventeen-year-old recent high school graduate, who was known by his peers to be really smart and was working as a porter in a local grocery store, enjoyed building rafts. On this particular day, Eugene was floating in Lake Michigan on a fourteen-by-nine-foot raft that four different groups of about twenty boys had worked on for almost two months. "It was a tremendous thing," one of his friends recalled years later to a reporter. But the fun was short lived. Eugene had accidentally crossed an invisible line and drifted into an area of Lake Michigan that white people had commandeered for themselves. Angered by the perceived intrusion, a white man began throwing rocks at the teenager, stoning Eugene as he tried desperately to swim away. He drowned right there in the Great Lake. His murder was ruled an accidental drowning. Local white press outlets described him as a teenager who died after "invading" the white beach. His death sparked a wave of violent riots committed by young white gangs who viewed Blacks in the city as an unwelcome sight. Their attacks throughout the city resulted in the deaths of twenty-three African Americans and fifteen whites. There were millions of dollars of property damage as one thousand Black families were left homeless as a result of arsonists.

Conservatives love to cite Chicago's violence as a coded way to shade Black people as incapable of living peacefully in our own communities. Trump exploited the death of former NBA player Dwyane Wade's cousin, Nykea Aldridge, tweeting in 2016 about Chicago's gun violence. "Dwayne Wade's cousin was just shot and killed walking her baby in Chicago," he tweeted, misspelling Wade's name and forgoing offering sympathy to the family. "Just what I have been saying. African-Americans will VOTE TRUMP!" Jeff Sessions, as attorney general, tried to blame the violence on Chicago's sanctuary city policy, saying it protects "criminal aliens who prey on their own residents." But it was white violence that first destroyed the city. Racist policies instituted by racist politicians helped bolster multifaceted inequalities throughout these municipalities, creating innumerable barriers for Black people to rebuild in the process. It's either willful ignorance or an intent to rewrite history that affords these ill-advised, obtuse talking heads the temerity to misspeak, spreading such a gross misrepresentation. A lie. Either way, once people of color become the collective majority in the United States, there will be reeducation camps for all white supremacists and their direct descendants. (Obviously not really—but just thought I'd write something to make the few racists who might actually read this book completely panic and explode. After all, Fox News has to have something for privileged pinheads like Tucker Carlson and Laura Ingraham to blabber about incoherently.)

The cruelest and bloodiest of the violence took place in Elaine, Arkansas, a small town on the Mississippi River. Sharecroppers there were being shaken down by landowners. Each season, white landowners demanded obscene percentages of the profits, they refused to give the sharecroppers a detailed accounting, and they trapped them with cooked-up debts. The workers had to purchase everything at the plantation commissary, which hiked up prices. This included food, clothing, tools, seed, and fertilizer. Black people were set up to forever be financially imprisoned by their debt. They thought organizing and joining a union was the only way to survive. It was the evening of September 30 when the sharecroppers and their families gathered in the Hoop Spur church to discuss joining the Progressive Farmers and Household Union. They thought this would help them secure a fair price for the cotton they picked—which in 1919 was the most profitable in history—

and they stood to pocket a decent profit. They aimed to buy the land they worked and had planned to hire a white lawyer to represent them with the landlords. But news of their plans leaked, and those who held all the power at that time were never going to allow the playing field to be leveled.

At around 11 p.m., a group of local white men, some of them likely affiliated with local law enforcement, fired shots into the church. With their wives and children inside the house of the Lord, Black men were not going to just lay down and die. They returned fire, and a white man was killed. Word spread rapidly, and rumors of an organized "insurrection" against white residents spread across town. The governor at the time, Charles Brough, called for five hundred soldiers from nearby Camp Pike to round up the "heavily armed negroes." The troops were under order to "shoot to kill any negro who re-fused to surrender immediately," according to reports in the *Arkansas Democrat*. But no one was ever given an opportunity to surrender, which of course would have meant death. The soldiers were joined by local white vigilantes and armed with a machine gun. For five days, they embarked on a killing spree, hunting Black people within a two hundred mile radius. Blood drunk, they burned homes with families inside. They savagely tortured any Black person they could get their hands on. They seemed to delight in inhumanity. Those they didn't kill were left maimed, scarred, and forever traumatized by the experi-ence. It was a massacre of American citizens by other Ameri-cans. Black life in the community was annihilated. By the time the chaos ended, the white mob had indiscriminately murdered more than two hundred African Americans—men, women, and children were all slaughtered. Not one white person was ever held accountable for any of those deaths. But someone would pay the price for the five white people who died. On October 31 of that year, a grand jury indicted 122 Black men and women for concocted offenses ranging from murder to night riding. Twelve of the Black men were convicted by a white jury for the murders of three white men, despite two of the deaths occurring from white people accidentally shooting each other. Let that sit for a while. Black patriotism is a motherfucker, ain't it? Today, Elaine is a majority-Black town where many of the residents live below the poverty line. They're represented in Congress by a white Trump Republican (kind of redundant) who, at the end of 2019, voted to oppose restoring parts of the Voting Rights Act.

Sadly, none of these stories are unique. Riots like these exploded all across the country. There was unrest in Houston, Texas; Indianapolis, Indiana; East St. Louis, Illinois; Washington, DC; Omaha, Nebraska; Charleston, South Carolina; and more. In every case, white people were the aggressors. And the deaths tore through Black households, ripping apart families, depressing their financial outlook, fracturing hope, and scorching paths to prosperity. Wives wept. Mothers wailed. Fathers faltered. Husbands agonized. Children cried. And far too many died. Just like the barbarity that preceded it, the impact of the Red Summer would ripple through time, impacting generations to come.

So as present-day atrocities happen—like the kidnapping and caging of Hispanic children at the southern border, racist comments and tweets from government officials, mass shootings carried out by white men with the intention of killing minorities, and so on—there's a chorus of mostly white people who decry the resurgence of overt racism by saying, "This is not who we are." But isn't it, though? America was not founded on peace and prosperity. It was taken from the Native Americans with blood and fury. It was not built on equity and equality. It was built on the backs of the enslaved. The country did not allow opportunity for everyone; it afforded power to a privileged few. Their American dream was our American nightmare. Black people lived in constant fear that the power of white people could, at any moment and without any notice, devastate their entire world. Let's not feign surprise when white supremacy rears its ugly head again and again and again. It's a system that is as American as apple pie and as old as the Battle of Jamestown.

———

Russian operatives understood the liability that white supremacy posed for the United States as far back as 1919. That America's greatest weakness was its hypocrisy: a nation founded on principles of liberty and justice for all, that enslaved Africans and massacred the indigenous. The early part of the twentieth century was a perfect time for them to exploit that weakness. That year marked the first time the Kremlin targeted a frustrated and disenfranchised Black community in America, exploiting the violent race relations of the early century that were oppressing large swaths of the country. Elected officials, including President Woodrow Wilson, feared Blacks returning from the war who were advocating for labor rights and other equalities. He stoked the racism by predicting that Black people "would be our greatest medium for conveying [Soviet communism] to America." It is hard to tally the exact number, but nearly one hundred African Americans were lynched in 1919. Yet the white people were the ones panicking. They feared the disenfranchised would overthrow the government to create a new regime modeled after the Soviets. The Red Summer bled into the Red Scare, a fear of anarchism and Bolshevism, the communist form of government adopted in the Soviet Union following the Bolshevik Revolution of 1917.

In the form of the Communist International, or Comintern, a report from Ozy found that the Soviets approved a $300,000 fund for propaganda targeting Black America. In today's numbers this would amount to roughly $4.5 million (recall that Russia spent $1.25 million per month on its campaign in 2016). In 1928, the Comintern started organizing workers and targeting Blacks in the south with propaganda showing pictures of Russia's former head of state, Vladimir Lenin, with captions like "LENIN Shows the South the Only Way to JOBS, LAND and FREEDOM." Several former Black Communists testified before Congress in the 1950s, when an FBI report, "The Communist Party and the Negro," was declassified, revealing Russia's true nefarious intentions of the 1920s. The Soviets had no intention of improving American society for Black people, according to one declassified FBI document. Their goal was "to exploit legitimate Negro grievances for the furtherance of communist aims." Russia has made a habit of routinely meddling in US elections. But their skill to sow racial discord would especially come in handy nearly one hundred years later.

Russia's interest in Black people during the early parts of the last century also piqued the interests of government officials at the time, who then missed the mark. Instead of focusing on the enemy across the ocean, they turned inward and focused on their perceived enemy right here in America: Black people. Any Black person who spoke up for racial equality was likely to be investigated by a network of federal intelligence agencies, as noted by historian Theodore Kornweibel Jr. in his book *Seeing Red: Federal Campaigns against Black Militancy, 1919–1925*. It didn't matter if these activists were actually sympathetic to communism. It was an excuse to surveil and ultimately control. When Black people resisted the tumultuous violence from white people during the Red Summer, investigators directed then Justice Department official J. Edgar Hoover to find out how those Black people had been radicalized. And when his agents in the field returned empty-handed, not having found any evidence of such radicalization, Hoover asked them to try again. And then again.

Weaponizing surveillance against racial justice activists became a habit. This practice was repeatedly used against Black people throughout the civil rights era in the 1950s and 60s, through COINTELPRO. The FBI conducted covert activities against Black leaders who were advocating for full equality, including Martin Luther King Jr., Malcolm X, Ella Baker, and the Black Panther Party. This type of discriminatory surveillance continues today. A 2017 FBI report disseminated to at least eighteen thousand law enforcement agencies nationwide asserts, without evidence, the existence of so-called "Black Identity Extremists Likely Motivated to Target Law Enforcement Officers." Documents suggest that the FBI had in recent years put resources toward running informants, as well as physical surveillance of activists. They also reveal that the agency employed open-source information, such as social media, to profile and track Black activists.

There was no social media back in 1919, so white press outlets filled the role of inflaming the idea that people in southern Black communities were being radicalized. Hence, the white supremacists in government incorrectly thought that any resistance to the abuses of Jim Crow might be evidence of foreign meddling. Again, they were more concerned with Black people than they were with the foreign meddling itself; they were just as dumb and racist then as they are now. The irrational fear gave way to the Palmer Raids, named for then attorney general Mitch Palmer. The government carried out raids in thirty-five cities, sweeping up thousands of immigrants and suspected radicals. There was no due process. No search warrants. People were grabbed off the street or pulled from their homes. They were often badly beaten by the police and thrown in jail, left to rot there for months. Scary how history repeats itself.

In 1919, under the blood-soaked awning of oppression, spirits buoyed. Those who were left standing did not retreat in fear. They had lost some battles, but they were determined to win the war. The dead had not fallen in vain. There were Black people who had survived the terror of the enemy. But before freedom could ring, more fighting would have to happen. Black people, having been robbed of their entire genetic identity, had long declared America to be their home—the only one they knew.

And African Americans were determined to make the country fulfill its promise. Though they knew it would not be fulfilled in their lifetime, perhaps their children would live to see a better day. The change they hoped for was down a very long road. And still . . . the journey continues.

6

~~White Economic Anxiety in West Virginia~~ Black Economic Anxiety in Michigan

I have friends who used to work for Michelle Obama. When the former First Lady started her sold-out book tour for *Becoming*, a ton of people began reaching out to all of them. Some people they hadn't spoken with in years, some were only casual acquaintances, and some they had never met at all. You guys know how this goes. The former White House staffers were getting the same random texts and DMs. "Hey girl! I see you out here killing it. So you know I hate to ask because I'm sure everybody is probably hitting you up right now but . . ." Everyone wanted tickets. Obama's book tour was the literary equivalent of a Beyoncé concert. My friends had all gone through some life changes over the years—everything from childbirth to sick parents. They hadn't heard from random people then. But for Michelle Obama's book tour, the randoms were boldly reaching out (I get it—we *were* talking FLOTUS after all). Well guys, that's how it can feel being a Black voter. We haven't heard from you in fifty-eleven Sundays, but then out of the blue, when you need us—our phones are blowing up. As part of the rising majority of America, I've got to say that this tactic will no longer work.

But hey, I can understand why some people foolishly think this is the way it works because so much of what people know about politics is filtered through the media. And despite Black people being loyal viewers of broadcast news—six in ten say this is their preferred pathway to news, according to Pew—very few political conversations focalize issues that uniquely relate to African Americans. For example . . .

Sidelining Black Voters

If I hear a reporter utter the phrase "white economic anxiety" one more time, I may slam my laptop against the wall and not finish this book. If you're reading this, praise be. That means I found another outlet for my frustration. During the campaign, the cable news outlets consistently aired a barrage of images of unruly white middle-aged men and women wearing MAGA hats gathered in auditoriums, physically harassing (and some assaulting) protestors. The media painted them as the working-class voices from Middle America expressing "white economic anxiety," thus birthing a myth about the type of people who were supporting Trump. This was a lie. Yes, unemployed West Virginia coal miners voted for Trump. But so too did a lot of middle- and upper-class white people across every geographic region, according to election night data. White working-class people didn't deliver the White House to Trump—white people everywhere did. So how did the media get this so wrong?

Newsrooms are filled with people who, intentionally or not, prioritize issues impacting white people. By 2032, a majority of the working class will be people of color, predicts the Economic Policy Institute. But still, the economic anxiety of other communities takes a backseat when it comes to that of white folks. I mean, what exactly is white economic anxiety? It sure must be a powerful enough ailment if it can drive so many white folks to ignore racism, xenophobia, misogyny, and sheer incompetence to elect a billionaire reality TV star parading as the champion of the working man to an office for which he was grossly unqualified. I presume this economic anxiety that is somehow exclusive to white people is the fear that people of color are doing well while white people are not? News flash: that's not the case. Never has been. Are those suffering from this mysterious "white economic anxiety" deranged enough to believe that people of color in America are ever the ones winning? At their expense?? Excuse me while I burst into laughter. And then tears.

There must be mental side effects to this epidemic. Given the history of this country when it comes to the economics broken down by race, ever wonder why the term "Black economic anxiety" was never coined? After all, Black voters certainly take our economics with us into the voting booth. Moreover, we've certainly impacted elections with measurable results. Yet I've never seen any such phrase attached to our financial stress—even though that stress is everywhere. Black economic anxiety isn't limited to the Rust Belt, or the Deep South, or to non-college-educated voters, or single moms, or to coal miners, factory workers, or farmers, or to cops, or bikers. The economic anxiety from which we suffer transcends geography, education level, home ownership, environment, marital status, and occupation. The economic anxiety we bear is both hereditary and airborne. And people in the state of Michigan know this all too well. In this battleground state, Black economic anxiety excludes you from democracy altogether, apparently. It was there that 75,000 Black voters literally didn't count in the 2016 presidential election.

Wait, WHAT? Yes, that's right. In 2016, over 75,000 ballots cast in Detroit—the most populous city with the largest Black population in America—were not counted in the presidential contest. You guys don't remember this? Maybe because it wasn't a breaking news story across every major network and barely got any coverage at all. I mean . . . it's not like it was a Trump rally or anything. If this had been a screwup in any other capacity, it would have blanketed the networks. If 75,000 people had their flights delayed at a major airport, that's a major breaking story. If 75,000 people couldn't get registered for Obamacare on the ACA website, it's a four-day news story. If 75,000 people have their data stolen from Target, it's talked about across new platforms for days. But 75,000 people in a big city in a crucial swing state being discarded from democracy, and barely a peep—likely because they're Black.

According to Michigan election officials, the 75,000 ballots were counted in down-ballot voting, but it could not be determined for whom those 75,000 people had voted for president. Hence the votes were tossed out. Trump won the state by 10,704 votes. Had Hillary Clinton received those 75,000 votes, she would have easily won the state of Michigan. One of the main reasons this began to get national press coverage is because 2016 presidential candidate Jill Stein challenged the election results in Wisconsin, Michigan, and Pennsylvania. Then both national and local news outlets began covering the recount, which was halted by the courts.

So how the hell did that happen? Many reasons, but basically it was white supremacy masquerading as the GOP. Republican state officials took direct control of government spending in Detroit. The result? Eroding the voting systems in the Blackest city in the state. More than eighty voting machines in Detroit malfunctioned on Election Day, according to officials. Paper ballots were collected but inconsistently recorded by the machines. The ballots in question were not eligible for a recount because of an obscure 1954 Michigan law that prevents recounting votes when there is a discrepancy in ballots. Democratic lawmakers introduced a number of bills that would have allowed no-reason absentee ballot voting before the 2016 election. This would have allowed the city's residents to send in ballots early, which would have given authorities advance notice of any mechanical malfunctions. But the Republicans who had controlled the state government since 2011, thanks in part to gerrymandering, shut those bills down like they were the old-school bad boys (the Pistons! See? I know sports sometimes).

The GOP that year was pushing for a restrictive in-person voter ID requirement instead. In November 2019, voters passed voting rights laws that allow people to register to vote on Election Day, vote absentee without providing a reason, and allow residents to automatically enroll to vote at the Department of Motor Vehicles. But since none of those laws were in place in 2016, especially early voting, there was no way to detect any issues, since the machines are used one day a year.

Despite all the talk about foreign adversaries meddling in US elections and less talk about domestic efforts to keep elections one-sided and unfair, Michigan—home to over one million Black people—was clearly the victim of both. Let's look at the economics for Black people in the purple state. Four years before the 2016 election, nearly one in five Black workers in Michigan were unemployed—that was more than twice the rate for white workers. The unemployment rate for Black people was 18.7 percent, compared to 7.5 percent for white workers in the fourth quarter of 2012, according to a report from the Economic Policy Institute. These numbers began to decline as 2016 approached. And by the end of 2017, the Economic Policy Institute found that the unemployment rate for Black people in Michigan was 10.8 percent. The state's overall unemployment rate in 2018 was 3.8 percent.

But if you have the impression that Black workers were getting hired, hold tight. Not necessarily. The sheer numbers alone don't tell the full story. The challenge with the way the media presents these numbers is that anchors and reporters rarely provide the proper context. The federal government defines the unemployment rate as the total number of unemployed as a

percentage of the labor force. People are considered to be unemployed if they aren't working, could work, and have searched
for a job within the past four weeks. However, people who aren't
working but haven't looked for a job within the past month
are considered to be out of the labor force. They're not counted
at all when it comes to official jobless statistics. Additionally,
though the national unemployment rate is the lowest in fifty
years—a stat often touted by media outlets—they rarely add
the footnote that this figure primarily reflects white employment dynamics. Not Black folks.

This is also not to declare that unemployment rates are not
improving. But this improvement certainly has not translated
into broader economic gains for Black workers in the state. The
racial wealth gap has worsened not only in Michigan but nationally as well. The median net worth of a white household was
twice that of a Black household in 2016. Not a lot of discussion about that in 2016, though, because the mainstream media
was too busy carrying Trump rallies live and unpacking *white*
economic anxiety. Also, while unemployment is one factor, un-
deremployment is another. For instance, temporary workers
across the American economy are disproportionately Black.
In 2018, temporary jobs supplied by temp agencies reached a
new high of 3.2 million in the United States. So while Black
workers constitute 12.1 percent of the overall workforce, they
make up 25.9 percent of temporary agency workers. And temp
workers made up a significant portion of those on the picket
lines during the 2019 General Motors strike in Detroit and
throughout Michigan, which offers some economic insight into
the area's Black voters.

Temporary workers account for 7 percent of GM's US
workforce, and they earn as little as $15 an hour at the start.
In Detroit, this doesn't get you very far. The city's population,
by design, is mostly poor and Black. A confluence of urban re-
newal and discriminatory loan policies keep Black people inside the city's impoverished core, which is stifled with blighted
buildings that are expensive to repair. Because of Detroit's prox-
imity to other major cities, corporations often opt for locating
just outside the city, which impacts the municipality's ability to
attract capital. So when the longest autoworkers' strike in fifty
years finally ended in the fall of 2019, there was a collective sigh
of relief for the nearly forty-eight thousand GM workers. But
certainly not a victory lap.

The strike began because workers were demanding increased
job security, a pathway for temporary workers to become permanent, better pay, and retention of healthcare benefits. Also,
GM had caused an uproar when the company announced plans
to close up to five factories in the United States and Canada
and cut more than six thousand jobs in the United States.
Leadership was attracted to the cheap labor in Mexico. Workers were striking to save an assembly line in Lordstown, Ohio,
and transmission plants in Warren, Michigan, and Baltimore,
Maryland. The striking workers were unable to save the three
plants. Globalization dominated the forces of collective bar-
gaining.

As a result of the deal, permanent workers received an
$11,000 signing bonus, and temporary workers got $4,500—
not really a win since it basically covered the wages they lost
during the strike. GM employees in Warren, Michigan, had

been collecting food, diapers, and baby formula to help striking workers get by. They were getting a $275 weekly stipend from the union—try surviving on that alone. When things are that bad, you take what you can get. But the thing of note about the strike was the folks on the picket line. Attention all political reporters regurgitating the phrase "white economic anxiety" in the echo chambers: did you all see what I saw? The footage and photos of the nearly fifty thousand striking workers revealed a clearly diverse bunch. I saw white, Black, and brown faces, women, young and older workers, families of different ethnicities, who all worked together at GM. These, of course, have always been the faces of the working class. But most were cloaked in color, rendering them invisible to the white media elites assigned to perpetuate two mythical tales of poverty: one for white folks—undeserved and confounding; the other for us—self-inflicted and to be expected.

This brings us to Flint, Michigan. Remember this majority-minority city? The one where the water was poisoning people, and the media paid attention to it for about two months back in 2016? Residents here may not fit the appealing national narrative of poor coal miners, but certainly what the people of Flint suffer could rival a Tyler Perry movie tragedy. The median household income here in 2016 was $26,330—over $30,000 less than the median household income for the country. Almost one in six of the city's homes has been abandoned. Nearly 42 percent of Flint residents live below the poverty line, with Black people making up the majority of this number. Money doesn't grow on trees, and safe drinking water doesn't flow from the taps.

The city's water crisis began in 2014, when, to save money, the city switched its drinking water supply from Detroit's system to the Flint River. Soon after, residents began complaining that the water from their taps looked, smelled, and tasted foul. City officials replied that the water was safe to drink; it was not. Inadequate treatment and testing of the water devastated the already struggling city. They had failed to add needed corrosion controls to the river water. Lead from the city's old pipes leached into the water, causing alarmingly high lead levels in the blood of many residents. Women were losing pregnancies. Babies bathing in the water were covered in blisters. People were losing their hair. There was a national, albeit brief, outcry from the media. But while they moved on, the city remained broken. While we sit through countless segments unpacking the needs and frustrations of Trump voters, poisoned water in a blighted American city has been normalized. The residents of Flint are still drowning in the problems this travesty created.

In the months following the election, what we heard throughout the media landscape is simply that Black voters stayed home. Some cable news pundits attributed it to Hillary Clinton–induced apathy. Some reporters blamed Russian troll farms. Some journalists offered no explanation and simply reported the data. But no one reported on the human experience of Black people. No deep-dive interviews with Black families in the state and the storms they were weathering. The Black experience was not unpacked with the same narrative afforded to white people. Much of the media merely presented a lazy narrative that Black people had simply opted to forgo exercising their civic engagement.

The Michigan government and the media failed the Black residents in this state. The many voters who did endure the well-placed challenges in the voting booth in 2016 were still discounted and denied their role in democracy. Did some Black voters stay home? Yes. But it was not only because of Russian bad actors. The GOP political machine was also determined to sideline Black voters. And Trump, with his crass, racist language, really excited the white folks in Michigan in 2016. The Center for American Progress found that the state's 2016 voters were 82 percent white—up 1.5 points since 2012, with both white college-educated and non-college-educated voters experiencing roughly equal increases in turnout. If we are to say that the collective of Black voters who abstained from voting did so because of Russia, we must also acknowledge that these voters did not cast ballots because of America as well.

James Baldwin said, "The future of the negro in this country is precisely as bright or as dark as the future of the country." Like it or not, our fates are intertwined. Ergo, given the state of America, it would behoove those in politics to not view Black voters as a necessary commodity—or a problem—during election cycles, but to instead view Black people as fellow citizens who are integral to the functioning of the country.

———

Now, if you will, picture a single man eyeing an ex-girlfriend who is paying him no attention. The more she ignores him, the more he tries. In his mind, she is simply playing hard to get. He is determined and thinks that at some point she won't be able to resist his charm and will come running back. Meanwhile, he's ignoring all the beautiful women around him completely open to be courted, while the ex is at the crib singing along to Beyoncé like, "Sorry, I ain't sorry." In this scenario, the single man is the Democratic Party. And who is the ex? White voters. They left you a long time ago, homie. They're not coming back.

Nationally, Democrats have not won the majority white vote in any election since 1964—the year President Lyndon Johnson ran, and he lost that year's white southern vote to conservative Barry Goldwater. So it begs the question: Why do Democrats who run for national office seem so eager to appeal to a demographic that abandoned them long ago? Moreover, why does the media perpetuate the myth that white voters determine Democratic politics? This puts Black voters in a position to have to scream and shout to remind Democrats who makes up their faithful base. The data is clear. The percentage of white voters needed to win the presidency continues to decline for Democrats and increases for Republicans, as the American voting electorate becomes more diverse every year.

Take President Barack Obama, for instance. In 2012, he received 5 million fewer white votes than he did in 2008. Yet he still prevailed over Republican Mitt Romney, who received one of the highest percentages of the white vote in the past forty years—59 percent. Hispanics and Asian American Pacific Islanders also moved toward Obama, manifesting their consolidation with the Democratic party.

The research shows that diversity, outside of national politics, has a clear partisan divide. A report from the Reflective Democracy campaign found that in 2019, people of color held

27 percent of Democratic seats at all levels of government, while less than 3 percent of Republican seats are held by non-white people. Yet a main talking point among some national Democratic candidates is their ability to attract "swing voters" and win in red states. Psssst—you're running as a Democrat. Talk to us about how you plan to win in blue states. Furthermore, let's debunk the idea of a "swing voter." Somewhere between 6.7 million and 9.2 million voters cast ballots for Obama in 2012 and then cast a ballot for Trump in 2016. This allowed some grossly out-of-touch talking heads from the chattering class to perpetuate the lie that Trump won by tapping into economic anxiety and not racial animus. As evidenced by the examples I gave in Michigan, we don't need research and data to call b.s. on this one. But for those who do, political science researchers at UCLA, UC Riverside, and Princeton took a statistical look at a large sample of Obama-Trump switchers. They found that these voters tended to score high on measures of racial hostility and xenophobia—and they were not likely to be suffering economically. So can we please stop perpetuating this false narrative? And while we're at it, stop referring to "could be possibly MAGA supporters" as "swing voters." Shorty, swing my way. Because if you're still on the fence in these extremely divisive partisan times, news flash: you are emphatically not a swing voter. People who cannot possibly be persuaded to break faith with a party that has exhibited or been an apologist for treason, xenophobia, sexism, racism, misogyny, pedophilia, abuse, and murder should never fall under that misnomer. Yet still, some candidates have defined their own electability as having the gift to reach the

MAGA crowd. Does the white vote just hit differently or something? Guys, make it make sense to me.

The entire election system is becoming anachronistic, as evidenced by the overstated importance of Iowa (a state that is 90 percent white) and New Hampshire (a state that is 93.2 percent white), the first two states in the country to hold primaries. And look how great that turned out in 2020? Not only did Iowa bungle the process by using a hastily built, insufficiently tested, flawed smartphone app, but the state's Democratic Party was unable to release the results until days after caucusers voted. It didn't stop members of the media from tripping over themselves to cover the state though they rarely put it in context. Iowa only accounts for forty-one of the Democrats' 3,979 pledged delegates. New Hampshire is not much better—they have a mere twenty-four delegates. The caucus and primary inflates a small number of people in these states with an insane amount of power in the electoral process. After both states delivered wins to 2020 Presidential candidate Bernie Sanders, the media declared him to be the frontrunner. They simply forgot that the voters, Black people and other voters of color, who determine the democratic nominee had not yet weighed in.

So why are these two states first? "It's mostly an accident of history, having to do with party reforms made between the 1968 and 1972 elections that took the decision of who would be the nominees out of the smoke-filled rooms and into the voting booth, giving us the primary system we have today," writes the *Washington Post*'s Paul Waldman. "Once the power they were wielding became clear, people in Iowa and New Hampshire made sure we all accepted that their status is somehow natural and proper. It isn't going to change; both states have laws mandating that their contests take place before any other states'." Nevada and South Carolina, which have a sizable share of people of color voting in the primary, voted third and fourth. But first, the candidates had to get through the first two primaries to be competitive.

The media would have you believe that these two states are a litmus test for electability. Both national and local reporters essentially drag candidates who don't perform well in the states, treating those who come in second or third as either the election season's biggest loser or a surprise victor. But watch Black voters shape democracy. Remember what came next? No dis to the good folks of New Hampshire or Iowa, but fried butter sticks or mayonnaise face carvings do not determine who will be the next leader of the free world. Voters of color do. The truth is, we're just talking about a few thousand votes in those first two states. When viewing the overwhelming attention paid to Iowa and New Hampshire, African Americans who may not engage in the political process may inaccurately feel like their vote doesn't matter as much, when it actually matters the most in many parts of the country.

When Louisiana's governor John Bel Edwards, the only Democratic governor in the Deep South, was aiming to pull off a second-term victory in a runoff election, he eventually looked to Black voters. To be victorious, he would need 90 percent of the Black vote and 33 percent of the white vote. Edwards was in a runoff because of what some analysts call a weaker-than-expected turnout during the first go-round among Black voters, who comprise about 33 percent of the state's population. Edwards in 2017 had signed into law the most comprehensive criminal justice reform in the state's history to reduce racial disparities, allowing Louisiana to shed its status as the state with the nation's highest imprisonment rate by the end of 2018. The bipartisan package of ten bills primarily focused on nonviolent, nonsex offenders and is designed to steer less serious offenders away from prison, strengthen alternatives to imprisonment, reduce prison terms for those who can be safely supervised in the community, and remove barriers to successful reentry. Among the reforms was reversing a rule around split-jury verdicts—a post-Reconstruction-era rule explicitly intended to disempower Black jurors and allow convictions without unanimous decisions. Many of these reforms stood to yield positive results for Black people, but there was no one telling that story to the masses of African Americans in the state, which Trump carried by twenty points in 2016. Between September and November, Edwards had thirty-four days to entice the community he had not been actively engaging. He rehauled his campaign, bringing on additional staff and consultants, which helped generate enthusiasm around his candidacy, and he was able to pull off a victory thanks to Louisiana's Black voters.

While LaTosha Brown, cofounder of Black Voters Matter, celebrates the victory, she cautions about efforts that look like this. "As long as the infrastructure is based on a model where it's centered around a candidate, when the candidate isn't running, all of that work disappears," says Brown. "When we talk about building democracy, I think this is the reason why it's really important that you're engaging people in the process—not just so you can win but so that people can feel a sense of their own agency. Because when people are not engaged in it, I question where you can sustain the win." She says when something is so centered around a candidate, then it's the candidate's victory and not the people's victory. "It's beyond just the transaction of convincing people to vote for you. You have to transform people to fight for change and to keep it."

Sustaining a win means expanding the base, making democracy as inclusive as possible. Republicans benefit from smaller voting bases because the smaller the turnout, the better their chances at victory. However, Brown says that is not something that's specific to the GOP. In fact, she says it's not something that is specific to white candidates. "No one wants to say this, but we see Black Democratic incumbents not expanding the base," she says. "They're not bringing in a lot of new voters because new voters mean younger voters, who tend not to be as loyal to the incumbent as their base. So you're not seeing an expansion of democracy." Brown says this applies to good candidates who are only interested in moving their base because ultimately that's what's going to help them win. That's fundamentally not what democracy is. Brown states that often for those in power, their first consideration is how can they protect *their* position of power. That has not led and does not lead to the expansion of democracy for Black people. Democracy doesn't say the victor is whoever has the best base. Democracy says that everybody should participate in the process. Activists like her have had to leapfrog over traditional media—that has long overlooked so many non-white communities—so she can help inform them.

Brown's organization treats communities as candidates, which makes sense. Why shouldn't communities have political strategists? That is how you create transformative change at all levels—not just in policy. She says she is building a political wheel that creates a culture shift, because something formed from a groundswell is difficult to unravel. "As racist as America is, you would be very hard pressed to find people—other than a few white supremacists—who would think it was okay to have a white water fountain and a Black water fountain. They may be racist, but even that's too much for them. Why? Because something changed. There was a culture shift that accompanied the political shift in this country, and that level of segregation was no longer acceptable. See? Those movements are more sustainable." But that culture shift, so far, has not reached get-out-the-vote efforts on a large scale when it comes to the way politicians have maneuvered within Black communities. Democrats running for local, state, or national office consistently try to win Black votes while still aiming to convince others that they're not beholden to Black voters. We see you.

In Virginia, Democratic governor Ralph Northam was mired in scandal since a 1984 photo surfaced of him in blackface, which appeared on his medical-school yearbook page. Amid national calls for his resignation, Northam maintained continued support from Black voters in Virginia, who made up roughly 20 percent of the electorate in his 2017 election against Republican Ed Gillespie. Do these voters really just love Northam so much that they're willing to blindly forgive his racist picture? Hell no. But most Black people are rarely surprised when something racist surfaces from a white person's past. Disappointed? Perhaps. Angry? Maybe. But surprised? Hardly ever. Even newly woke allies can be suspect. Don't get me wrong. We're here for the alliance, but fools we are not. Were Black people to summarily kick the governor to the curb, who would then be left? The two people in succession were ensnared in their own battles. The Black lieutenant governor, Justin Fairfax, had been accused of sexual assault by two different women in incidents from fifteen years ago. And despite there not being a criminal investigation, it was still a vulnerability. He has since announced a 2021 run for governor. And Attorney General Mark Herring had admitted wearing blackface at a 1980 party, further plunging the state into chaos as the trio of leaders were plagued by scandal. Black voters had few options.

If the top three Democrats were bypassed for leadership, the choices would have grown increasingly dim, as the next person in line for governor was a conservative Republican. Black voters, betrayed, really had nowhere to turn. Nor did Northam—he desperately needed Black voters. What a perfect time for Black voters to enforce a culture shift and detail *their* needs to him. If he has any hope of a political future, he would need to cater to Black political demands: affordable housing, maternal mortality, equitable funding of HBCUs, the removal of Confederate monuments, transportation equity, and more. The pragmatic Black voters of Virginia will begrudgingly ride with him until, that is, a more palatable leader surfaces. Writer Theodore Johnson describes it as "the chance to make tangible, incremental gains with an imperfect politician is preferable to exacting harsh political and social sanctions to prove a point about the unacceptability of past racist behavior, particularly if black interests could be further harmed as a result." It's far from an ideal scenario.

There are always tough choices for Black voters. African Americans are navigating a system that has historically kept them away from the ballot box. A 2018 Northern Illinois University study found that the Commonwealth was the second-worst state in the country on voting access, just ahead of Mississippi. But, with respect, Virginia has a lot of competition in terms of worst states for suppressing Black voters.

7

Georgia Outkasts

Nowhere is the strength of Black voters both flexed and flummoxed as it is in the state of Georgia. In a state where the Black, Asian American Pacific Islander, and Latino populations are growing faster than anywhere else, Georgia has the potential to become a battleground state. Yet coverage of the state's massive voter suppression was almost nonexistent until the 2018 election. It's not that Georgia doesn't get coverage. When the Super Bowl comes to town or Tyler Perry opens a massive studio, the press is all over Atlanta. But when secretary of state Brian Kemp has African American senior citizens pulled off a bus for trying to vote? Raiding voter organizing offices like some Dirty South gestapo? Not a peep. It's like Outkast said: the whole world loves it when you're in the news, the whole world loves it when you sing the blues.

Georgia's media market was ill equipped to handle what was happening in the state. "The state's largest paper, the *Atlanta*

Journal-Constitution, doesn't have any full-time African American political commentators or journalists," says Dr. Jason Johnson, who covered the race extensively when he worked for *The Root*. "This meant that issues dealing with Black voters didn't get the attention they deserved. One of the only ways we found out about the voter suppression Brian Kemp was committing was because of reporters like Darren Sands at *Buzzfeed*, Vann Newkirk with *The Atlantic*, and Errin Haines with the Associated Press." Atlanta's one Black newspaper, the *Atlanta Black Star*, just didn't have the capacity to cover these issues. And national markets, not focused on the rising majority populating progressive pockets in the south, tend to dismiss the Peach State as solidly red. But Georgia is becoming a critical state in the country's shifting democracy.

Democrat Stacey Abrams was the prototype. She ran a historic race for governor in 2018, winning more votes than any other Democrat to run for statewide office in the state's history. Abrams would have potentially served as the state's—and the country's—first African American woman governor. But when Brian Kemp ran in 2018 as an immigrant-snatching, abortion-ending, antibusiness, good ole boy, thousands of Georgians said alright alright alright! But none more than white women. They helped deliver Kemp directly to the governor's mansion, having outperformed white men voting for him at a 75 percent clip. In Jackson County, Kemp won over 60 percent of the vote. So if you're part of the 50 percent who support abortion rights, I'm sorry, Miss Jackson—Kemp was for real. He went after reproductive rights with a vengeance, almost leading to a nationwide boycott of Georgia, costing millions of dollars in film and television production, and leading other large industries to think about bypassing the Peach State in favor of an environment that's a bit less retrograde.

Under the guise of Christian evangelicalism, southern white women make up the core of female opposition to abortion—something Kemp made a campaign promise. Also, notice how these descendants of colonizers commandeered the term "pro-life," and then everyone in the media began identifying them by that misnomer as well? Even that term was colonized. "Pro-life" did not originally have anything to do with abortion; it was a term used by some in the early 1960s counterculture who were advocating humanity for children, those in the prison system, and homosexuals. Yet today, the "pro-life" crew, who claim to care so much about the lives of children, are pretty indifferent or silent about migrant children suffering at the US southern border or Black children gunned down by police. These "Christian evangelicals" must be crumblin' erb to make logic out of being pro certain fetuses and callous toward certain lives, and then declare themselves in favor of life in general.

Kemp signed into law a bill that would prohibit most abortions once doctors can detect a fetal heartbeat, which occurs around six weeks of pregnancy. Current Georgia law restricts abortions after twenty weeks of pregnancy. A federal judge has blocked the law from going into effect, as opponents argue it violates a woman's constitutional right to abortion as established by the US Supreme Court ruling in *Roe v. Wade*. The case will likely make its way through the courts—where Donald Trump and Senate majority leader Mitch McConnell have stacked nearly two hundred conservative justices reshaping the judiciary—and it could potentially land in the Supreme Court. Black voter suppression, a southern tradition that still flourishes in Georgia, has wide-reaching impacts. Keeping Black voters from the ballot box could upend reproductive rights for women in Georgia and potentially throughout the country.

Kemp was serving as Georgia's secretary of state at the time of the election; he presided over the state's elections, voting machines, and vote tallying. He was essentially serving as a referee in a game in which he was also playing. And his supporters literally wore their racism on their chest. In a photo from the campaign trail, Kemp gave a smile and thumbs-up gesture with Jim Stachowiak—a racist weapon-wielding terrorist and former police officer who had been previously charged with disorderly conduct. In the photo, Stachowiak was wearing a yellow shirt stating in large letters, "Allah is not God and Mohammed is not his prophet," along with a Trump hat and a Kemp sticker. Shake it like a polaroid picture. "This guy was part of an organization that stood outside a Stacey Abrams event with signs, threatening people, threatening violence against anyone who stood against Confederate statues and flags, and Brian Kemp took a picture with him," says Dr. Jason Johnson. "Kemp is an anachronism, a throwback. There is no governor in the United States of America who is more out of sync with the growing changes in his state than Brian Kemp."

It's true. Despite the ever-increasing population of people of color in the state, Kemp made his bigoted political positions clear in a series of racist campaign ads. In one, Kemp says, "I got a big truck just in case I need to round up criminal illegals and take 'em home myself." In another, he stares defiantly into the camera and proclaims, "I strongly support President Trump, our troops, and ironclad borders, and I stand for our national anthem. If any of this offends you, then I'm not your guy." To make sure he was Georgia's guy, Kemp executed innumerable dirty tricks backed by the power of his office. Some tricks he had used before—like creating organized noise.

During the 2014 midterms, when Abrams was the minority leader in the Georgia House of Representatives, she launched the New Georgia Project. The group's aim was to increase voter registration among people of color, young adults, and unmarried women. They were targeting more than seven hundred thousand unregistered Georgians. Leaked audio from a Republican breakfast captured Kemp saying, "You know the Democrats are working hard, and all these stories about them, you know, registering all these minority voters that are out there and others that are sitting on the sidelines, if they can do that, they can win these elections in November." Weeks later, Kemp accused the organization of voter fraud. In a subpoena, he demanded that the voter group turn over massive portions of its internal records relating to the campaign. "We're just not going to put up with fraud," Kemp said in an interview with a local news station at the time. But his own team had to admit they never found a thing. "We have not detected from anything that [the group's leaders] have said or done that it is a goal of the New Georgia Project to go out and commit voter registration fraud," Kemp's lead investigator said, according to an opinion blog in the *Atlanta Journal-Constitution*.

Undeterred, Kemp forged ahead with his office's criminal investigation, which turned up fifty-three allegedly forged voter applications—though state law requires canvassers to turn in applications regardless of what people write on them. "If the form says Mickey Mouse registered in Anaheim, California, we have to turn that form in," Abrams said to *Newsweek* in 2014. Still, Kemp called on Georgia's attorney general to prosecute the group for felony voter fraud, which carries up to ten years in prison and a $100,000 fine. Although Kemp's investigation cleared the New Georgia Project of any wrongdoing in February 2014, he waited until late September to announce it publicly. What's crueler than being cruel? Ice cold.

As secretary of state, he canceled more than 1.4 million voter registrations—nearly 670,000 of which were canceled just one year before the gubernatorial election. His office had 53,000 voter registrations placed on hold under the state's "exact match" policy, which he helped push through the state legislature in 2017. This rule says that a voter registration application is incomplete if information on that form does not exactly match records kept by Georgia's Department of Driver Services or the Social Security Administration. Reasons for no matches included the transposition of a letter or number, deletion or addition of a hyphen or apostrophe, accidental entry of an extra character or space, or use of a familiar name like "Tom" instead of "Thomas." Those who were flagged were disproportionately Black—70 percent. Four days before the 2018 election, a US District Court judge ruled that the exact-match policy presented a "severe burden" for voters. More than three thousand new citizens whose registrations had been held up were allowed to vote. But the burden remained for the bulk of Black voters.

Kemp was running for governor but couldn't fulfill his responsibility as secretary of state and run a proper election. Kennesaw State University, which runs the state's election system, in 2017 quietly destroyed a computer server shortly after a lawsuit was filed by voting rights advocates seeking to force the state to overhaul its election technology. Remember Michigan's 75,000 voters being discounted from democracy and no one being held accountable? It was the same in Georgia. "Do you know what would happen if there was a server at Georgia Tech that lost a year's worth of incoming freshman grades or financial data? People would be in jail for that," Dr. Johnson says. "You would have parents from all over the country saying, 'What the heck happened to my student's financial aid information? What happened to my kids' grades?' If that happened in a high school, in a hospital, if that much data was erased anywhere under any circumstances, heads would roll. So I think people would be shocked to know that, you know, you can make those kinds of mistakes or have that kind of corruption as a government agency and ain't nobody going to jail for it." It gets worse.

Hush that fuss, everybody move to the back of the bus. On the first day of early voting, organizer LaTosha Brown arranged for a bus full of Black senior citizens to go to a voting center in majority-Black Jefferson County when local officials removed them all from the bus. That's right—they literally pulled them off the bus like they broke the law. It wasn't like they had just wiped a computer server clean or anything. *They were going to vote.* Jefferson County administrator Adam Brett defended the move, stating that the bus had not been authorized to take the seniors to the polls and that the action constituted "political activity" at a county-run senior center. And while Kemp, by his own admission, was so concerned about minority voters making their way to the polls, the state continued to be overcome by innumerable challenges to those trying to cast ballots.

On Election Day 2018, all the players came from far and wide. Voters in predominantly Black counties waited at polling places for hours while hundreds of available voting machines sat unused in government warehouses. State officials had sequestered those machines in response to a federal lawsuit that said they were vulnerable to hacking. A cybersecurity expert had found in 2016 that he was able to access personal information about voters on Georgia's election system, as well as passwords for county election supervisors. One polling location at Atlanta's Pittman Park Recreation Center, a predominantly Black area, had just three machines. Local election officials had to extend voting hours in some locations to compensate for the problems. During Kemp's reign as secretary of state, county and state officials had closed more than two hundred polling places between 2012 and 2018. Some touch screens switched voters' selections, and ballots would "self-cast," according to dozens of affidavits submitted by voters. "Repeatedly had ballot changed from Abrams to Kemp," read one summary of a voter's affidavit submitted in February 2019. "Took three tries before it remained on Abrams." There was also the curious case of approximately eighty thousand people who did not vote for lieutenant governor but voted in other down-ballot races. Republican Geoff Duncan beat Democrat Sarah Riggs Amico by 123,000 votes in that race. Machines in at least four voting locations in Gwinnett County temporarily went down, causing voters to resort to paper ballots, according to the *Atlanta Journal-Constitution*.

A number of Black voters reported problems when they tried to vote, with some saying they were turned away due to the exact-match policy. Others cited problems casting absentee and provisional ballots. Kemp had long used the power of his office and the whims of conservative media to suppress Black voters.

When Brooks County, Georgia, elected a majority-Black school board for the first time in 2010, Kemp sent investigators to interrogate Black residents in the county about their unprecedented use of absentee ballots in the election. Law enforcement officials, at Kemp's direction, had registration activist Nancy Dennard arrested. She was taken from her home in handcuffs, placed in a squad car, and marched into the police station. Eleven of her political allies were also arrested and charged with 120 separate felonies. Among the arrestees were three newly elected Black school board members, who were stripped of their positions because of the allegations. The Quitman 10 + 2, as they came to be known, had their mugshots, all of them wearing orange jumpsuits, blanketed across the media as evidence of voter fraud. Their images were splashed across the front pages of newspapers, played on a loop on local TV news, and shown on Fox News as a talking point to prove voter fraud was real. The intent was not just to suppress but also to humiliate. In fact, Kemp has routinely sent the Georgia Bureau of Investigations to raid the offices of voter registration groups on trumped-up charges. Then Georgia Republican governor Nathan Deal removed the newly elected Black members from the Brooks County School Board in January 2012, two months after the indictments were filed. Dennard had been elected president of the board but was stripped of that title as well. As the court case against them moved forward, Dennard won reelection in the fall of 2012, along with the other African American member of the school board, Frank Thomas, and in early 2013 Dennard, Troutman, and Diane Thomas were reinstated according to the law. They served together for twenty-two months. Four years after the election, the accused were acquitted of all charges, but their lives had been devastated. And Black voters there decided civic engagement was not worth the vicious white backlash. In 2014, not long after their term, the board switched back to majority white.

In 2012, attorney Helen Ho was working to register newly naturalized immigrants to vote in the southeast. She discovered that some new citizens in her group, then known as the Asian American Legal Advocacy Center (AALAC), who had tried to register in Georgia, were still not on the rolls. Early voting had already begun, and some polling places were turning away new citizens seeking to vote for the first time. She penned a sharply worded open letter demanding that Georgia take immediate action to ensure the new citizens could vote. Kemp responded by launching a two-and-a-half-year investigation of the group and asked that AALAC turn over certain records of its registration efforts, citing "potential legal concerns surrounding AALAC's photocopying and public disclosure of voter registration applications." The investigation did not focus on voter fraud, but instead honed in on more technical issues, such as whether canvassers had people's explicit, written consent to photocopy their registration forms before mailing the originals to the elections office. "Our genuine desire was to help the secretary of state clear these people through to vote, so it was interesting that their response was to investigate us," Ho told the *New Republic* in 2015. "I'm not going to lie: I was shocked, I was scared." This group was also ultimately cleared of any violations.

Voter suppression is a hugely important story, but broadcast newscasts largely brush it off. Actually, the industry doesn't just blow off stories of voter suppression; they only look at the issue through the lens of whether the race is competitive. If they don't consider it competitive, they just bake voter suppression into the cake. But that's not how journalism is supposed to be done. It's not about exposing truths that pay into preconceived narratives. It's about informing the public. The *Tyndall Report*, which monitors newscasts, found that during the 2018 midterms, the three major networks— ABC, NBC, and CBS—together aired fewer than ten voting-rights stories between Labor Day and Election Day. Considering that more than five million Americans watch the evening news on network television, a good chunk of America is likely unaware of the threat voter suppression poses to democracy. Because the issue mostly impacts Black people and other communities of color, America is barely aware that the problem exists at all. According to a 2018 Pew Research report, among likely voters, 84 percent expect that voting will be easy (with 44 percent saying that it will be "very easy"), while only a small minority of respondents (15 percent) believe that voting will be "at least somewhat difficult." Overlooking this story aids in the disenfranchisement of Black people. What happened in Georgia is not an anomaly. The Brian Kemps of America have long been the norm and not the outcasts.

8

Ohio and the Purge

There's an annual holiday in dystopian America that falls just after Halloween, when misconduct that *should* be illegal is celebrated by a small few. On this day millions of unsuspecting Americans gather to decide how society should function. But lurking in the statehouse are culprits who crafted a devious, socioeconomically unbalanced plan designed to keep the master malefactors in power. The miscreants do not bear masks like in the movies. Yet their trickery helps them transform transgressions into law. The holiday is known as Election Day to most of us. But to the unmasked marauders it is known as the Purge. Let's travel to Ohio, where I was born.

For many years Ohio has been considered a battleground. Ohio has voted for the winner in every presidential election since 1964. The Midwestern Rust Belt has been described as a microcosm of America. From the urban streets of Cleveland to the suburban cul de sacs of Columbus, it is thought to encompass most of the demographics once sought after by political strategists. But . . . does it?

In actuality, the state is no longer the snapshot of the country. But the media has not caught on yet. For one, Ohio is *not* as racially diverse as the rest of the country. It's not so much that people of color have left the state. It's that people of color are not flocking to it. In fact, *no one* seems to be heading to Ohio. Its population growth is lagging—the state ranked forty-eighth of all the states from the year 2000 to 2013, with just a 1.9 percent increase in new residents. The remaining population, of which more than 81 percent is white, is aging, and the birth rate is declining. Thanks to a sluggish economy, it's projected that Ohio will barely surpass 11.7 million people between 2020 and 2030. So . . . why all the hoopla over this state? The obsession with Ohio by the press and by candidates is driven by that same white notion about which values and which voters are more important. Ohio has an ideological and cultural resonance to white media elites, and it's time for it to die. It used to be important, and the romanticizing of Ohio diners and truck stops may have made sense to some in the 1970s, but those cultural touchstones aren't nearly as relevant today.

Ohio's celebration of the purge is speeding up its fate of being considered a lot less politically relevant in national politics. In the movie *The Purge*, the upper class allows a twelve-hour period when all crimes—including murder—are legal. The reason? To rid society of the undesirables. And that's exactly what Ohio is doing with its voter rolls—attempting to purge those they consider to be "undesirable" voters: mainly Black people. Since 1994, Ohio has had a policy of purging infrequent voters from the rolls. The state is allowed to cull voters' names if they haven't voted in six years and don't respond to a postcard informing them that they are about to be purged. The standards for which a voter can be purged are exceptionally low and disproportionately impact lower-income and Black households, where people are less likely to vote regularly. So basically droves of Black people must go through the burdensome process of jumping through arbitrary hoops just to vote. The GOP is attempting to reduce the value of certain citizens to their voting frequency—something that is absolutely contrary to how Democracy functions. After all, the Republicans who champion these draconian rules might have to purge their own beloved leader, Donald Trump, if these are the guidelines for being able to vote. According to the *Washington Post*, Trump only voted in eighteen of twenty-eight general elections during the period 1987–2014. Should *he* be purged?

In 2012, Ohio had about 8 million registered voters. Officials sent out 1.5 million notices that voters were set to be purged, and 1 million were left unanswered. But how many people closely examine a postcard arriving via snail mail that blends perfectly with the countless pieces of junk mail that flood our mailboxes every day? I'd imagine not a lot. A Reuters analysis found that in Ohio, "voters have been struck from the rolls in Democratic-leaning neighborhoods at roughly twice the rate as in Republican neighborhoods." In 2015, a whopping forty thousand voters from Cuyahoga County (Ohio's largest county, which includes the city of Cleveland and a significant number of Black voters) were purged—many of whom were in fact eligible to vote. The purge is intentionally partisan and unfair.

President Barack Obama won Ohio in 2008 and again in 2012. In 2016, the state went to Donald Trump. But just because Ohio has picked previous winners doesn't mean it will continue to be a political crystal ball. Somebody better alert the political analysts and their outdated maps ASAP. "Ohio has been dying as a state of electoral importance for three cycles now," Dr. Johnson says, after having lived there for a number of years. "Had Hillary Clinton won Arizona and Wisconsin, she would have won the presidency. She didn't need Ohio. And when the new census comes out, Ohio is going to have the same amount of electoral votes as North Carolina. Ohio is just not that important. LeBron isn't there anymore." NBA superstars aside, the state is definitely losing its status as a battleground showdown. Ohio has been trending toward Republican candidates for years. This is no accident; this political trend is by design. In order for the purge to be successful—meaning they had to make sure droves of Black people wouldn't make a fuss about being eliminated from participating in democracy—the GOP had to make the process of voting as difficult and inconvenient as possible. And the mainstream press ignored it all.

With many Black voters living in predominantly Black communities across the country, Election Day experiences are often riddled with challenges. *The Nation* reported that during the 2004 election, Ohio had the longest lines to vote in the country, with five-hour waits in heavily Democratic cities like Cleveland and Columbus. They also found that a postelection report for the Democratic National Committee estimated that 3 percent of Ohioans—174,000 people—left their polling places without casting a ballot. Pollster and Democratic strategist Cornell Belcher states that Black voters waited an average of fifty-two minutes to vote, while the wait for white voters was only eighteen minutes. Moreover, he noted that twice as many Black voters reported experiencing problems at the polls in 2004. To address these challenges, Ohio adopted thirty-five days of early voting. Some portions of that chunk of time came to be known as "Golden Week," when Ohioans could register and vote on the same day. This was widely used during the 2008 election—the year Obama was elected, with Ohio playing a role in his decisive victory. In the 2012 general election, more than 45,000 Cuyahoga County residents voted early in person—and 77 percent of them were African American. According to state Democrats, Black voters were also almost twenty times more likely to use early in-person voting than white voters.

The Grand Ole Party had just the plan. In February 2014, Ohio's Republican-controlled legislature passed a law that killed Golden Week. Days later, Ohio's then secretary of state John Husted announced a ban on Sunday voting. A federal judge later required Husted to restore early voting on the weekend and Monday directly before the election. But the court stopped there. Ohio's then governor—Republican John Kasich, who was up for reelection that year—signed the voter suppression bill into law. Kasich, who can be seen regularly on CNN often decrying what a travesty Trump has been for the nation, seems painfully unaware of his role in helping to deliver Don the Con directly to the White House—considering that Trump won Ohio in 2016 thanks to the rules Kasich helped put in place. He has not been confronted about his role in this blatant act of voter suppression. This is where media literacy, diversity, and democracy again come into play. Despite having a permanent seat on the airwaves, Kasich has never been challenged on his issues with race. Ohio is home to some of the country's strictest voting laws. Kasich signed many of these bills into law, and for the ones he didn't (voter ID), he repeatedly refused to voice opposition. So he can miss us with those tears now. The scene of Trump taking a big crap all over democracy is one that Kasich (and so many others like him) helped write, direct, and star in.

So in an effort to purge Ohio's undesirable Black voters, the GOP has set the state on a course of political irrelevance. Their multipronged voter suppression efforts have created a deluge of problems in the state that overwhelmingly and negatively impact its Black residents. There really is no happy ending in *The Purge*, nor does there appear to be one for Ohio. It continues to move further away from being a battleground state and closer to being a Republican stronghold, despite the political views of the people in the Buckeye state. Because of the state's long-standing position as a perceived geographic indicator of the nation's political direction, there's a common saying among politicos aiming to reflect the state's power. Unfortunately, the same phrase can be applied to the state's relentless efforts to silence its Black voters: So goes Ohio, so goes the country.

9

Florida Man

Are you familiar with Florida Man? No? Okay, to explain, we need to visit *Atlanta*. Not the city, but the brilliant FX television show executive produced by performer Donald Glover. He portrays Earn, and while driving with his conspiracy-theory-citing friend Darius (the talented Lakeith Stanfield), he says that his parents are heading to Florida. Darius warns Earn to tell his parents not succumb to the legend of "Florida Man," who, he explains, is responsible for a high percentage of abnormal incidents that occur in the state. "Think of him as an alt-right Johnny Appleseed. No one knows his true identity, date of birth, what he looks like. That's why headlines always say Florida Man," Darius explains. Over a nightmarish and hallucinogenic montage of a faceless white man committing deranged crimes across Florida, Darius cites real-life Florida Man headlines as examples to make his point: "Florida Man Shoots Unarmed Black Teenager." "Florida Man Bursts into Ex's Delivery Room and Fights New Boyfriend as She's Giving Birth." "Florida Man Steals a Car and Goes to Checkers." "Florida Man Beats a Flamingo to Death." "Florida Man Found Eating Another Man's Face." At the end he asserts that Florida Man and the state government are in cahoots "to prevent Black people from coming to and/or registering to vote in Florida." Did Glover nail it or did he *nail* it?

Now, am I trying to say that this stream of nameless mugshots of crazy demented criminals in Florida is part of some larger plot by politicians to sideline Black voters in the state? No, of course not. The guys I'm talking about have names, and none, as far as I can tell, are demented. They're actually *very* smart and strategic. Even smarter than the media, as their misdeeds were able to go undetected for an entire campaign cycle. Let's start with the current governor and former Republican congressman Florida Man Ron DeSantis. He rode the Trump train to national prominence specifically in right-wing circles.

Then congressman DeSantis was facing off with Democrat Andrew Gillum, former mayor of Tallahassee, in the 2018 Florida gubernatorial race. In one of DeSantis's on-air political ads (appearing to be fresh from a *Saturday Night Live* sketch), things got super weird. DeSantis's wife appears on screen saying, "Everyone knows my husband, Ron DeSantis, is endorsed by President Trump, but he's also an amazing dad." The next scene is DeSantis handing toy blocks to his toddler daughter and encouraging her to "build the wall." Then DeSantis is seen reading to his younger son, Mason, while quoting Trump's former reality show, *The Apprentice*. "Then Mr. Trump said, 'You're fired,'" DeSantis says while holding his infant son in one arm and Trump's book *The Art of the Deal* in the other. (Side note: *The Art of the Deal* is the 1987 book coauthored by journalist Tony Schwartz, for which Trump takes credit, and where Trump offers—get this!—business advice.) Next, DeSantis is seen teaching his daughter how to read while spelling out "Make America Great Again," on a Trump campaign poster. "People say Ron's 'all Trump,' but he's so much more," his wife says sarcastically, as Ron checks on their son, who is in a crib wearing a red "Make America Great Again" onesie. Where was his headline: *Caught on Tape: Florida Man Passes On Lessons of White Supremacy to His Two Young Children*? This warped *Children of the Corn* political ad actually resonated with the other Florida Men (and women). Could DeSantis be part of a deep-rooted conspiracy to keep Black people in Florida from voting? Hell yes!

I had known of Gillum, who was thirty-seven years old at the time he ran for governor, for some time. The world is small for the contingent of Black folks who attended HBCUs, and smaller for those who work in politics and policy. This familiarity, of course, is not considered valuable in the incestuous world of white media elites. Their close-knit circles of PWI (predominately white institution) alumni, or colleagues who dated friends, or all those crews who are habitually on the same invite lists to cocktail parties or bar crawls seem to matter more than the relationships organically honed by those of us navigating both worlds. Which is why those insular circles of the chattering class, who still dominate media coverage, so often miss the mark. Just as they did with Gillum.

Florida's gubernatorial race is always national news given

its status as a battleground state. Gillum's first opponent was former congresswoman Gwen Graham in the Democratic primary. This daughter of a former governor and senator was the presumed Democratic frontrunner. As such, Gillum's campaign was summarily dismissed by all those political experts who decorate cable news sets. "This is Graham's primary to lose," the pundits said repeatedly in unison. But certainly reporters on the ground in Florida saw the crowds he was commanding. Surely these cable news pundits had bothered to check in with actual Florida voters before declaring the primary Graham's to lose. Let me be more clear: Had they bothered to speak with any Black voters on the ground? Gillum, who had been on a short list of leaked names in 2016 as a consideration to be the running mate of Hillary Clinton, never seemed to fall on the radar of the white Beltway talking heads. Until, that is, August 28, 2018. Much to their surprise (not ours), he clenched the Democratic nomination for governor of Florida. It was described as a political upset. But upset to who? The out-of-touch aristocracy being celebrated by anchors and executives who enjoyed hearing their exclusive group's opinions as they espoused political analysis from yesteryear?

Gillum lost his bid to DeSantis despite showing an impressive ground strategy of engaging voters and not letting the media define his electability. He ran a campaign that was in stark contrast to the values of Trump's GOP. He backed the Medicare for All healthcare plan, supported impeachment, and wanted to replace US Immigration and Customs Enforcement. But DeSantis had something Gillum did not: the affection of white supremacists and their beloved leader, Donald J. Trump, who campaigned hard for DeSantis. Come to find out, there was a good reason Florida Man was so blatantly enamored with Trump. DeSantis had long kept the company of racists and conspiracists. Yet this was not immediately reported by major Florida news outlets. Months after DeSantis entered the race, it was reported that he previously spoke at the David Horowitz Freedom Center Conference in Palm Beach, Florida, and Charleston, South Carolina. Not to be confused with a MAGA rally, this conference is also known for its leader's anti-immigrant and anti-Muslim ideologies. Horowitz had previously referred to Palestinians as Nazis, openly opposed reparations for descendants of the enslaved, and claimed that the only "serious race war" in America is "against whites." DeSantis spoke at this conference four times: in 2013, 2015, 2016, and 2017. Some of the organization's regular speakers include former Trump presidential advisor Steve Bannon; pedophilia apologist and alt-right commentator Milo Yiannopoulis; and the Dutch far-right leader Geert Wilders. Florida Man was excited to join this fray. "I just want to say what an honor it's been to be here to speak. I've been to these conferences in the past but I've been a big admirer of an organization that shoots straight, tells the American people the truth and is standing up for the right thing." That's how he greeted conference attendees in 2015, according to the *Washington Post*.

DeSantis also chose not to return money from a donor who called President Barack Obama a "fucking Muslim nigger." And in August 2018, the day after winning his primary, DeSantis appeared on Fox News, where he warned voters that his African American opponent, Gillum, would "monkey this up" if he won the election. Before that, one of his top fundraisers had been suspended from Fox News for asking a Black panelist if he was "out of his cotton-picking mind." Around this time, it was also revealed that DeSantis was the moderator of a racist, Islamophobic, and anti-immigrant Facebook group called "Tea Party." Just days after this revelation, Gillum became the target of robocalls by a neo-Nazi group.

My point in this book is how media coverage enables the erasure of Black folk from democracy. By informing the public, the media can sway public opinion. In most cases. But this case may be an outlier. In Florida, I must be honest about the limitations of the media. Given some of the actions DeSantis had taken that were covered in the media, his supporters could not have been surprised. Had there been more of a national outcry that yet another Trump-like character linked to bigoted groups and espousing racist lingo was going to be the executive of a state, perhaps a few Floridians might have switched their vote. *Perhaps.* But for the most part, the people in Florida knew who DeSantis was. None of this hurt him. In the Sunshine State, it helped him. The people of Florida elected to give him dangerous and unyielding power to shape a huge part of democracy in his racist vision. As soon as he took office, he began to undermine the will of the people, starting with an amendment to restore voting rights to over a million people.

Florida was one of four remaining states (along with Iowa, Kentucky, and Virginia) that still barred people convicted of felonies from voting after they returned to society. Nationally, 6.1 million people are disenfranchised due to felony convictions. About 40 percent of this group is African American. This means one of every thirteen Black people cannot vote due to voter disenfranchisement. In Florida, nearly 65 percent of the people voted in favor of Amendment 4—a ballot initiative that would restore voting rights to those who had served their time, except for those convicted of felony sexual offenses or murder. The law had been on the books since 1868, when lawmakers, forced by the federal government to allow Black people to vote, had devised it as a way to prevent African Americans from gaining power. This was done in tandem with post–Civil War laws that made it nearly effortless to incarcerate Black men. Before Amendment 4, those with felony records who wanted to vote again had to petition a state parole board for the opportunity. This process could sometimes take years, and the outcome varied depending on the current governor. The state routinely rejected thousands of applications, leaving a disproportionate number of Black people unable to vote.

So now we were to believe that the same state which had elected Trump as president and Trump acolyte DeSantis as its next governor had also passed a new policy that would add an estimated 1.4 million voters to the rolls? And one-fifth of these voters would be Black people? Since not too many supporters of the Republican Party would be in this new wave of voters, this could certainly turn the purple state of Florida more blue. Something had to be afoot. There's no way Florida Man was going to abandon his quest to keep Black people from voting just like that. A barrier to this influx of voters was inevitable. I'd put money on it.

Well, turns out money was precisely the thing that was the barrier. At DeSantis's urging, Republican lawmakers in the state voted to allow only those who had paid off all the fees, fines, and surcharges attached to their sentences to become eligible to vote. So let's get this straight. Black folk, saddled with unreasonable debt assigned to them by a racist system, must buy their way out from under racist policies designed specifically to make it easier to for them to be arrested and kept out of the voting booth? And we're not talking about a $20 fee here. This unconstitutional poll tax was likely going to be considerable, given that Florida is a national leader in racking up financial penalties, which have become a top source of revenue for the state. A local Florida news outlet reported that more than $1 billion in felony fines were issued between 2013 and 2018, according to reports from a statewide group. Florida assesses more than 115 different fines against defendants. Court charges unpaid after ninety days are handed over to private debt collectors, who can add a surcharge of up to 40 percent to the fines and fees incurred by the convicted.

How many Black people who had been convicted of a felony crime could afford to pay these monstrous fees? But that was likely Florida Man's intention all along. The Brennan Center for Justice found that in the first few months after Amendment 4 went into effect, there was a significant surge in voter registration, much of it driven by Black people reenfranchised under the new law. The group also found that the newly registered voters reported incomes far lower than the state average. The racial implications are clear: this financial caveat would decrease the number of Black voters considerably. And this was all by design. In February 2020, a federal appeals court ruled that the legislature's effort to limit voting to only felons who have paid off all their court debts was unconstitutional. So the Florida governor, having been dealt a blow by the federal courts, turned to "monkey up" the judiciary in his own state.

Governor DeSantis is reshaping Florida courts with lots of help from the Federalist Society, a powerful nationwide organization of conservative lawyers that is also helping Trump pick justices. Florida Man's friends used to wear white robes; now they wear black ones. After DeSantis was elected, he got to replace three State Supreme Court justices who had been appointed by Democratic governors, remaking Florida's high court into one of the most conservative in the country. He appointed one of his own lawyers (a Federalist Society chapter president who had virtually no experience in the courtroom) as chief judge of the Division of Administrative Hearings—whose twenty-nine judges function as the public's check on state agencies and boards. Florida's Supreme Court Judicial Nominating Commission, stocked with Federalist Society members, did not recommend a single Black person to the bench, leaving the high court without a Black justice for the first time in decades. The conservatives in Florida want to keep the ballot box as white as they now keep their courts.

It certainly does seem like the state is in cahoots with Florida Man to prevent Black people from coming to or registering to vote in Florida.

10

You Down with GOP?

I'm going to say it: nearly every Black Republican I have met has seemed slightly off to me. For starters, the party is overwhelmingly white, and it's only getting whiter. There is no ethnic voting bloc that supports the GOP in staggering numbers, which is why Republicans choose to focus on courting white voters. They also routinely support cuts to programs that benefit poor people, who are disproportionately African American. And the party wittingly touts racist legislative actions and rhetoric. So how many Black people see a group of mostly racist white people and say, "Oh yeah, that's my tribe right there." According to Pew Research, 8 percent—that's how many Black people are registered Republicans. Yet, looking at cable news, there is a disproportionate number of Black Republican commentators relative to the number of Black people who are allowed to say anything at all on television. These GOP talking heads may have Black faces, but their voices—which bizarrely regurgitate racist conservative talking points—do not represent any large swath of the Black community.

To be fair, there's always an exception to this "*every* Black Republican is weird" rule. After all, I'm sure if I thought about it long enough I could come up with a few. General Colin Powell, for instance. And, ummm . . . yeah, that's about it. But even the more rational Black Republican(s) are not necessarily people we would expect to sit across from at a Spades table or see doing the electric slide at a Black wedding reception—well, save for former RNC chair Michael Steele. But for the most part, the oxymoronic Black Republican is more than a little peculiar. And when you take this tiny group of peculiar people and then parse out the Trump supporters—well now we've got a full-on minstrel show. Yet they are the cable news networks' version of a high-speed car chase: maybe bookers think viewers won't be able to look away, secretly hoping to see a crash and burn.

Let's start with the dumb girl that goes on the dumb network and says a bunch of dumb things. What's her name? Ah, yes. Candace Owens. Hardcore Trump supporters have elevated *her* as the Voice of Reason to encourage droves of Black people to exit the Democratic Party. This new younger and wackier version of Omarosa (we didn't think it could get worse, until it did) found a sweet spot after she completely reversed course from running a website that mocked Trump (including his penis size, I might add) to becoming a hard right-wing conservative gifted in stroking Trump's shriveled little orange . . . ego. This stark about-face happened within the space of a year. Owens was the CEO of Degree 180, an anti-Trump, liberal-leaning website, which in 2015 featured such takes as "good news" that the "Republican Tea Party . . . will eventually die off (peacefully in their sleep, we hope)," according to *Buzzfeed*. Back then, Owens talked about Trump being a racist with an immigrant wife. But by the end of 2016, the site shut down, and Owens magically "came out" as a conservative on YouTube a few months later.

Personally, I don't care about this slow-witted, self-hating souvenir of the GOP whom a bunch of white Republicans have elevated to stardom for their own nefarious purposes. While some hip-hop artist may choose to dignify her self-serving, ill-informed, empty, and ridiculous outlook on African Americans and our voting habits—have at it, homies—I haven't bothered to spend time trying to figure out why. Hip-hop mogul Diddy gifted her a platform to spew her gibberish in 2019 for a Revolt TV discussion on Black issues (Black issues?? She should definitely be on somebody's couch talking about *her* issues. But damn sure not one aimed at uplifting the plight of Black folks). And socialist, activist, and rapper Killer Mike calls her a "sister." The same woman who pranced up to Capitol Hill for a congressional hearing (at the behest of Republicans on the committee) to declare that white supremacy and white nationalism are no longer issues. The groundswell she got from white conservatives is paying off. For now.

But again, what is her political experience? This administrative assistant turned blogger turned GOP mouthpiece gets to debate accomplished attorney Monique Pressley about . . . anything?? Granted, this is usually on Fox News, but the issue is not specific to any network. Many white bookers and producers make no distinction between the lack of experience and expertise among this crowd. You can be an ill-informed Black Republican with zero political experience and be gifted a platform on any of the cable news outlets. But Black Democrats must meet an inexplicably high standard with extensive experience and policy knowledge and be able to do TV extraordinarily well. *The Root*'s Michael Harriot reviewed an entire year of NBC's *Meet the Press* and looked up the education level of every person who appeared on the show. He tweeted the results. "Some never went to college. Some had bachelor's degrees. But for an ENTIRE YEAR, EVERY SINGLE BLACK PERSON on MTP had a master's degree or a PhD except 1. And she had a little thing I like to call the Pulitzer Prize. And that, dear friends, is white supremacy." Harriot also pointed out that Chuck Todd, the show's host, who infamously said in 2019 that he was naive about Republican dishonesty, never graduated from college. This is not to say that not having a degree should preclude anyone from weighing in or leading political conversations. But do we think that I, a Black woman with no letters behind my name, would *ever* be considered to host a network's flagship political show?

Moreover, the new crop of Black Republican voices has no agenda other than to antagonize. Michael Steele was actively trying to recruit Black people to the Republican party during his time as chair. That's not what these GOP neophytes, who are essentially Trump lackeys, are trying to do. Despite being taken seriously by cable news bookers, they seem more like victims delirious with Stockholm syndrome. Who are they trying to convince or inform? Not us. Black people are not so easily swayed by skinfolk who come bearing the messages of those hell-bent on disenfranchising our kinfolk. A 2019 AP-NORC poll found that only 4 percent of Black people said they think Trump's actions have been good for African Americans. One of these 4 percenters is Houston native Jono Thomas, who attended the second annual Black Leadership Summit hosted by the White House in 2019 in his Sunday best to bow down to the MAGA leader: a top hat, white dress shirt, gray pinstripe pants, and wool scarf (in the summer!!). Again, not a lot of normal Black Republicans out there. But standby for this guy to make his TV debut any day now.

Also, the change in GOP outreach has been clear to anyone on the ground, but the coverage in the mainstream media has stayed the same. Much of Trump-era outreach to the Black community has been aimed at sidelining Black voters—not a true effort to envelop African Americans into a conservative movement. Let's not forget, Trump's campaign tactics mirrored those of the Russians. In fact, a senior Trump campaign official told Bloomberg reporters in 2016 that they had "three major voter suppression operations under way," including one aimed at depressing Black turnout. During his victory tour after the election, Trump actually praised Black voters for staying home. "We did great with the African American community," he said. "So good. Remember—remember the famous line, because I talk about crime, I talk about lack of education, I talk about no jobs. And I'd say, what the hell do you have to lose? Right? It's true. And they're smart, and they picked up on it like you wouldn't believe. And you know what else? They didn't come out to vote for Hillary. They didn't come out. And that was a big—so thank you to the African American community."

Leading up to the 2020 election, the MAGA crew teased out efforts at engagement with no clear strategy. They were not investing any real money in states heavily populated by Black people. Their entire game plan was to hype the false narrative that Black voters were abandoning or should abandon the Democratic Party. What they lacked in measurable evidence they made up for in embarrassingly foolish anecdotes. But the numbers don't lie. There has been no significant jump in support for Republican candidates from Black voters.

Yet white anchors and reporters consistently center much of their campaign reporting around white Republicans. The consistent question to candidates running at the federal level is "How will you appeal to swing voters/Trump voters/Midwest voters/working-class voters" or whatever other euphemism they want to use for white members of the GOP. It's a strange obsession the media has with Trump Republicans. They routinely invite on his hardcore supporters and parade them as swayable voters—which they clearly are not. Viewers suffer through these internet conspiracy theorists trying to clumsily talk through policy, about which they are painfully ignorant. It's baffling because I don't recall that kind of anthropological case study on Obama voters across the cable news media landscape. After all, they too disrupted the political terrain and toppled a political dynasty. And they did it without yelling racial epithets and xenophobic rhetoric at campaign rallies. Were they never interesting enough for a cable news panel?

The New Deal and an Old Problem

The drastic transformation of the party of Lincoln and Emancipation rivals the metamorphosis of 2004 Kanye rapping, "They make us hate our self and love they wealth," into 2018 Kanye rocking a MAGA hat. Emotionally tormented hip-hop artists aside, Black people were not *always* reliable Democratic voters. Republicans love to point this out. It is true: the GOP was founded on anti-enslavement principles. But there's not a remnant left of that party today. Back in the late 1800s, many Republicans believed in welcoming immigrants with open arms. And many in the party supported Black suffrage. Not that Black folks were easing our way into voting booths back then, but most African Americans did in fact identify with the Republican Party from Reconstruction until the election of Franklin Roosevelt in 1932. In fact, every African American who served in the US Senate belonged to the GOP—that is, until Carol Moseley Braun's 1992 election. And in the lower chamber, the House of Representatives, twenty-one Black men served exclusively as Republicans before a Black Democrat was elected. But the migration of Black folks from the Republican Party aligns with the increasingly anti-Black transformation of the party itself. It starts with the New Deal and an old problem.

The Great Depression, which lasted from 1929 to 1939, was the worst economic downturn in the industrialized world. The old cliché "When America catches a cold, Black America catches the flu" proved to be most true during this time. Few suffered during this period more than African Americans. At this time, Americans got their news from radio, film newsreels, magazines, and newspapers. And the only documentation of Black suffering was in Black-owned newspapers, which were protesting segregation, mistreatment, and discrimination. Black workers were already greatly depressed before the stock market collapsed in October 1929. Still, only white economic anxiety was acknowledged—especially then. Thousands of sharecroppers had already lost agricultural jobs in the mid-1920s due to the declining cotton market. They were already in the grips of an economic noose. Black people were the first to see hours and jobs cut, citing last in, first out policies. The African American unemployment rate in 1932 climbed to approximately 50 percent, according to the Library of Congress. In the south, it reached 70 percent. Though the day-to-day situation was already unimaginably bad, there was an added cruelty to the suffering of Black people: in many public assistance programs, African Americans frequently received substantially less aid than whites. Some community service organizations excluded Black people from their soup kitchens. Can you imagine? The inhumanity of that time, elevated by an economic crisis, rippled through the decades, still impacting Black people today.

During the Great Recession, which lasted only from 2007 to 2009 (thanks, Obama), the disproportionately large drop in African American wealth shows that Black people *still* encounter systemic obstacles in not only building wealth but maintaining it as well. For many Black voters, it has never been *just* the economy, stupid. Additionally, it has been our safety and the right to exist on equal footing and free of violence in the country built by our ancestors. But the country's economics have always impacted politics, and the intentionally brutal policies produced from it often disproportionately impact Black people in an unfavorable way. And the cable news outlets are frequently preoccupied with segments on the struggles of the white working class. Black economic suffering was happening on an even larger scale in the earlier part of the twentieth century.

Let's put what was happening with America's two major political parties in context: Imagine if Donald Trump suddenly declared himself a Democrat and President Barack Obama suddenly declared himself a Republican. But they each held dear their ideological views and brought them to their newfound respective political parties. The absurdity of this is akin to me suddenly declaring myself Beyoncé despite falling short in the areas of singing, dancing, and resisting the urge to eat carbs. But, in terms of our politics, that's essentially how the Republican/ Democrat realignment began to take shape back in the 1930s.

Republican President Herbert Hoover, who was in office at the beginning of the Great Depression, was the first president among many to discount the importance of Black people and the few Black voters who existed at the time. His administration failed to implement economic policies to help Black people, who had suffered more than most during the Great Depression. The Republicans had a faithful following of Black voters, who, for a time, remained loyal to Hoover. It didn't matter. Hoover was willing to ride with white southern segregationists. How did Black folks catch word of this? The Black press. According to the *Chicago Defender*, Hoover had spoken favorably of making the Republican Party "lily white" and had the covert backing of the Ku Klux Klan. How proud might he be of the GOP that exists today?

At that time, it was the Democratic Party that was suppressing the political rights of Black people in the south with fervent dedication. And most Blacks still lived in the south, where they were essentially prevented from voting at all. However, the Great Migration north was swelling with droves of Black people seeking a better life, and the tide was turning. This sea change was aided by Black press outlets, such as the *Chicago Defender* and the *Pittsburgh Courier*, which helped spread the word about what was taking place in the GOP. In fact, for decades the *Defender* used its coverage as activism to warn Blacks to flee the south. The goal was twofold: the preservation of Black life and hurting the southern economy. "The bars are being let down in the industrial world like never before in this country," read one of their editorials. "Unless we spread out to the four corners of the earth and accept every chance for advancement, we will continue to vegetate, as many drones in the southland are now doing." The paper not only advertised opportunities available in the cities of the north and west, but they also published first-person accounts of success. While the south responded with barbaric violence in a brutal attempt to force Black people to stay put, white-owned southern newspapers were playing a major role in racial violence. They ran enthusiastic stories about lynching. They served as cheerleaders for white supremacy, and their racist coverage had sometimes fatal consequences for African Americans. Black press outlets were the only counternarrative spreading lifesaving guidance about the geographical and political plight of Black people.

As the tide shifted for African Americans, so did the political parties. It wasn't that Black people left the Republican Party; rather, the Republican Party left Black people. So when it came time to elect a new president in 1932, the few Black voters bet on President Franklin D. Roosevelt. FDR beat Hoover in a landslide. It wasn't that FDR solicited Black voters on his first go-round. In fact, he was rather vague about his plans. He appeared happy to relish in the fact that Americans were so dispirited by Hoover's shortcomings as president that all he had to do was watch Hoover fall. This further proves the point that Black people are not loyal to any particular political party; Black people are loyal to ourselves. We would never put all our faith in a system that was designed without considering our well-being. For the sake of our own safety and our ability to survive and thrive, Black people have a history of maneuvering in the perilous spaces of whatever political party will be the path of least resistance to our own pursuit of happiness.

The choice has always been easy, but never simple. Some misguided right-wingers who have little engagement with actual Black voters perpetuate the disproven theory that African Americans overwhelmingly vote Democrat without putting any thought into the vote. Ha! How preposterous that a privileged few think that Black voters—whose abuse, torment, poverty, and deaths have mostly come at the hands of racist policies—can afford to not be keenly aware of who and what it is they're voting for. After centuries of navigating precarious racist terrain, Black people are often the first to recognize white supremacy. And as such, we are well aware that it is bipartisan. It penetrates every avenue of the nation's government.

Even allies carefully toe the line when it comes to recognizing and reconciling the injustices that took place in America at the hands of white people cloaked in the American flag. Which brings us back to FDR.

The thirty-second president of the United States strengthened Black support for the Democratic Party, in part by allowing Black leaders unprecedented access to the White House. For the first time, Black people in America had a direct line to their government. A select group of Black intellectuals was tapped to represent the interests of the very people who had been forced to build the nation that was still actively oppressing them. They came to be known as his "Black cabinet." Among them: William H. Hastie, who in 1937 became the first Black federal judge; Eugene K. Jones, executive secretary of the National Urban League; Robert Vann, editor of the *Pittsburgh Courier*; labor and civil rights activist A. Philip Randolph; economist Robert C. Weaver; and the sole woman at the table, educator Mary McLeod Bethune, who served as the National Youth Administration's director of Negro affairs.

Today, Roosevelt is praised for his New Deal, a series of programs and projects intended to offer relief to aid in the recovery of the national economy and benefit the unemployed. As he fundamentally changed the role of the US government as it relates to its citizens, he was inclusive of African Americans in his New Deal policies—largely thanks to his wife, Eleanor Roosevelt, and his Black cabinet. Policies included making low-cost public housing available to Black families; enabling Black youth to continue their education via the National Youth Administration and the Civilian Conservation Corps; giving jobs to many Black workers through the Works Progress Administration; and supporting the work of many Black authors, including Zora Neale Hurston, Melvin B. Tolson, Arna Bontemps, and Waters Turpin, through the Federal Writers Project. There was also a substantial shift in labor unions. The Congress of Industrial Organizations (CIO) for the first time organized large numbers of Black workers into unions. By the time Roosevelt was out of office, there were more than two hundred thousand African Americans in the CIO, many of them officers of union locals.

But like every relationship I had in my twenties, FDR's legacy is complicated. He knew that in attempting to get out of the economic quagmire that was the Great Depression, he was going to have to play chess, not checkers. But for Black people, particularly at that time, it's impossible to look at life as moving pieces on a board. It's never a game for the oppressed. Roosevelt had inherited a Congress occupied mostly by southern Democrats, who had been the face of oppression for Black people. Still, FDR acquiesced to them for the purposes of passing his New Deal agenda—which wasn't exactly the great equalizer for Black people.

Even under these new policies, whites still got the pick of jobs and were paid higher wages. The Federal Housing Program was essentially state-sponsored segregation. The affordable government-subsidized housing it created for the poor catered to whites, rejecting Black people from housing loans, and forcing these American citizens to remain in poorer neighborhoods. When measures were introduced to address this injustice, discrimination continued at the local level. The New

Deal included the Agricultural Adjustment Act, which paid white farmers not to farm in order to reduce crop surpluses and increase crop prices. But how would this benefit Black people who were sharecroppers, not owners of farmland? The wages they earned from sharecropping were determined by their white employers, who were the sole beneficiaries of all proceeds from the crops.

FDR also declined to advocate for passage of a federal anti-lynching law and remained aloof about civil rights for Black people. Efforts to ban the poll tax that prevented so many Black people from voting were also rebuffed. But still, where were Black voters to go at that time? Back to Republicans who had long since abandoned them? Were they to not participate in Democracy at all and leave both parties, neither of which was enthusiastic about leveling the playing field for African Americans? There were few choices for the architects of America to continue their role in building and shaping the government. One Black newspaper editorialist noted, "Now, if anybody thinks we ought to leave this Democratic ship and jump back into the Southern Republican skeleton and help put some meat on its bones, they have got some more thought coming. Brethren, we had too hard a time getting on this ship and we are going to stay, sink or swim." The migration to the Democratic Party had been sealed by a New Deal draped in an old problem—white supremacy. FDR was the Black voter's only option. By his 1936 reelection, 71 percent of Black voters—including hundreds of thousands who had never previously voted—supported Roosevelt, according to the Joint Center for Political and Economic Studies.

By the time the Second World War happened, Black people were *still* living in segregation under Jim Crow. When the war was over, Democratic president Harry Truman, a former segregationist who was the grandson of slave owners, in 1948 desegregated the military—just weeks before his reelection campaign, bowing to pressure from Black voters and at the risk of losing white voters. The bigots didn't have to panic. The order was not truly executed until 1963.

Republicans Were No Friend to Black People

The segregationists did, in fact, panic. At the 1948 Democratic National Committee Convention, thirty-five delegates from the Deep South walked out and formed the Dixiecrat Party. They elected as their new leader segregationist Strom Thurmond, who would never again identify as a Democrat—though at his death he *was* identified as the father of Essie Mae Washington-Williams, a Black woman who was the daughter of a sixteen-year-old maid working in Thurmond's father's house. But I digress.

In 1957, Republican President Dwight Eisenhower sent federal troops into Arkansas to desegregate Little Rock Central High School and help usher in the Little Rock Nine after angry white segregationists, women and men, prevented them from entering the institution. "Mob rule cannot be allowed to override the decisions of our courts," he declared in an address to the nation. Then in 1963, Democratic president John F. Kennedy did the same to federalize the Alabama National Guard and force desegregation at the University of Alabama. But the tipping point came in 1964: Lyndon B. Johnson signed the Civil Rights Act. Republicans would have you believe that the GOP supported the legislation and Democrats opposed it. Not quite. Nearly everyone from both sides of the aisle supported the Civil Rights Act—but not the south. It was southern politicians who were the dissidents.

Hearing about the atrocities committed by white people was more palatable for some reason than seeing them. In 1965, the power of Black people, white media gatekeepers, and violent white supremacy converged on Edmund Pettus Bridge. There was nothing new about this aggressive and violent bloodshed, which took place on the infamous voting rights march on Bloody Sunday—except that, this time, it was captured on national television. The dirty secrets of the south were no more; this revolution would be televised. The ruthless assault was brought into the homes of millions of people all over the country. ABC actually interrupted its telecast of *Judgment at Nuremburg*, a film about the prosecution of Nazi war criminals, to show what was happening in Selma, Alabama.

White rage, under the guise of law enforcement, had erupted. Inflamed troopers, mad with power and furor, attacked the marchers viciously. The group was brutally beaten with nightsticks, choked with tear gas, slapped with whips after they refused to turn back, and beaten bloody. White troopers on horseback chased down fleeing Black women and children. Civil rights icon and Georgia congressman John Lewis was beaten so badly that he suffered a skull fracture. It was a one-sided, bloody melee. It gave white people everywhere an up-close view of what Black people had endured for centuries:

terror. The horrifying images forced many to shed their previous indifference. Over the next forty-eight hours, demonstrations were held in over eighty cities in support of the marchers and voting rights for Black people. On August 6, 1965, President Lyndon B. Johnson signed the Voting Rights Act into law.

Black people were not simply fighting for the right to vote. "We were fighting for justice. We were fighting for relief. We were fighting for recognition of our humanity," says LaTosha Brown. "Voting was just one of the tools, one of the outcomes. But we were like no, as a human being I have a right to be seen to be heard. We have shifted the narrative that the civil rights movement was about us participating in their system and not about standing," she says. "In the midst of being beat down and attacked by dogs—we were supposed to love America? No, the civil rights movement was not about loving the county, it was about loving us. We can't put the value on the system and marginalize us." It has never been about the system. Systems can be changed.

By this point, lines had been drawn in the sand. Republicans were no friends to Black people. Now, here we find ourselves after nearly a century of Democrats having a stronghold on Black voters. This is not to say there are no significant challenges within the Democratic Party when it comes to our ever-crucial voting bloc. For starters, despite many of the 23 million Black women in this country being the most loyal voting bloc, the party seems to be more interested in attracting back white male voters who left the party long ago. Black women critical of the Democratic Party also say they are exhausted by being asked to knock on doors or write checks without being considered viable as candidates at the local, state, or federal level. Not as a DNC delegate. Black women also have significant economic power (African American women spend an estimated $1.5 trillion annually, according to Nielsen). Imagine tapping into those figures to elevate politics and policies that speak to our community. Paraphrasing the words of rapper Rick Ross: f*ck with us, you know we got it. Well, that's what Republican senator Tim Scott of South Carolina has tried to do.

Yes, the sole Black Republican in the upper chamber is trying to recruit more candidates of color to the GOP. Okay, so in the "I've never met a normal Black Republican" category, let's examine Scott. He has said emphatically that Donald Trump is *not* a racist, calling his comments only "racially insensitive." Scott also declined to join the Congressional Black Caucus back in 2011, when he was first elected to the House of Representatives. It was then that Scott cosponsored a welfare reform bill that would deny food stamps to families whose incomes were lowered to the point of eligibility because a family member was participating in a labor strike. But there are two sides to every token.

In 2016, when there were multiple stories of police brutality against Black people, Scott took to the Senate floor to offer personal testimony about his own encounters with police. "I've experienced it myself," Scott said. He revealed that he had been stopped seven times in the course of one year as an elected official—including once when the office thought his car was stolen. "Was I speeding sometimes? Sure. But the vast majority of the time I was pulled over for driving a new car in the wrong neighborhood or something else just as trivial." He ended with a plea to his colleagues to "recognize that just because you do not feel the pain, the anguish of another, does not mean it does not exist." He also opposed the confirmation of Thomas Farr, Donald Trump's nominee for a US District Court seat, due to unanswered questions regarding how much Farr knew about a decades-old effort to disenfranchise Black voters in North Carolina.

So he's not the inspiration for Samuel L. Jackson's character in *Django Unchained*, like some of these embarrassing oppressor-loving masochists. But the fact remains that he still rides with Trump. So that's why his efforts to recruit any Black candidate to run on a GOP ticket warrant an obligatory side-eye. This recruitment effort is being done in partnership with Jimmy Kemp of the Jack Kemp Foundation; Scott is honorary chairman of the group. While this might be a more serious action than trotting out clueless actresses (Trump supporter Stacey Dash), reformed shoplifter-turned-opportunist (Trump campaigner Katrina Pierson), and a confused talking head who appears to be going through an identity crisis (Trump syco-phant Paris Dennard), it still falls short of any real appeal to Black candidates who actually care about appealing to Black voters. Which prompts another question: Is this an effort to add a Black or brown face to a party grossly lacking any di-versity? Or do they genuinely want Black and brown voices at the table? With Scott hitting the streets saying Trump's policy agenda is a message that could appeal to minority voters, I've got a request: show us one. While Scott may be well inten-tioned, I contend that he is radically misguided. Perhaps he has been watching one too many cable news segments featuring Black faces from his ilk.

As history has shown, it's not the party; it's the policies. And the current GOP's policy priorities are largely informed by white nationalists whom the media refuses to deem such. Stephen Miller, for example, a speechwriter and senior adviser to Donald Trump, was the architect behind the Trump administration's hateful, draconian border and immigration policies and the author of much of Trump's hateful anti-immigrant rally rhetoric. Miller's repertoire includes the travel ban aimed at keeping Muslim people out of the country, separation of migrant children from their parents at the border, deportation of protected migrants and blocking refugees from entering the country, plus the relentless effort to limit asylum seekers at the southern border. He found a modern-day version of the white southern news outlets that advocated bigotry: *Breitbart News*, a far-right syndicated outlet offering content for bigots. A November 2019 analysis of more than nine hundred emails from Miller to editors at the site revealed in great detail Miller's unapologetic racism.

While still an aide to GOP senator Jeff Sessions of Alabama in 2015, Miller sent a series of emails to former *Breitbart* editor Katie McHugh (who, at the time, identified as a white nationalist). McHugh says Miller directed her to stories from the website American Renaissance—a white nationalist publication focused on eugenics and anti-Black racism. The website's founder, Jared Taylor, has argued that "when blacks are left entirely to their own devices, Western civilization—any kind of civilization—disappears." There's more. In 2013, he argued for a white ethno-state. "We want a homeland where we are a majority," he said. Insert fear of being a minority here. After

Dylann Roof murdered nine Black parishioners at a church in Charleston, South Carolina, in 2015, Miller emailed McHugh an angry message about retailers pulling Confederate flags from their stores. "What do the vandals say to the people fighting and dying overseas in uniform right now who are carrying on a seventh or eighth generation of military service in their families, stretching back to our founding?"

There are no published emails showing Miller's outrage when Trump attacked Gold Star family member Khizr Khan, the Muslim father of a slain US soldier. Trump, who has no military experience, questioned the Khans' motives for speaking out against him and mocked their religion in TV appearances and on Twitter, asking whether Ghazala Khan, the mother of the fallen soldier, "wasn't allowed to have anything to say." Her response: speaking about her dead son is too painful for her to do in public. There was also no outrage from Miller when Trump, an accused draft dodger who claims that bone spurs prevented him from serving, phoned the African American widow of army sergeant La David Johnson and couldn't remember her husband's name. But back to the emails he did send.

Miller sent McHugh a story from VDARE, a white nationalist website named for Virginia Dare, the first English child born in North America. This site is ridiculously preoccupied with "white genocide"—the myth that nonwhites are working to eliminate white people through immigration and interracial marriage. This is the same platform Miller cited in response to McHugh's concern that the government would grant temporary protected status (TPS) to Mexican survivors of the 2015 Hurricane Patricia.

"This being the worst hurricane ever recorded, what are the chances it wreaks destruction on Mexico and drives a mass migration to the US border?" asked McHugh. Miller replied: "100 percent. And they will all get TPS. And all the ones here will get TPS too. That needs to be the weekend's BIG story. TPS is everything." He accompanied his response with a link from VDARE that focused on the prospect of protected status for victims of the hurricane. Not a single Republican called for the ouster of a white nationalist driving policy in the White House. But why would they? Most surrendered their party to this type of despicable ideology a long time ago.

There are still a few Republicans who disparage the Trump era, which has warped and wilted their beloved GOP. But while they are disgusted at what has become of their Grand Ole Party, they don't seem to recognize the role they had in its shaping. Remember that, as Dr. Micheal Eric Dyson says, Frankenstein was the doctor, not the monster. Donald Trump is the monster they helped to create. And now that their party is cannibalized, some of these "Never Trumpers" want to build alliances with us? With Black people? In the words of the great urban philosopher and hip-hop artist Drake, "No new friends." We started from the bottom and now we're here. I remember well some of these people, particularly a few from the Black wing of the GOP. I remember their weird-ass "blue lives matter" tweets at the height of media coverage of police shootings. I remember them campaigning with and for Ben Carson, whose wokeness is as sleepy as his eyes. I remember the criticisms they unfairly unleashed at Obama.

At this critical moment in history, we *definitely* can't trust white folks who cheered on everything from Reaganomics to Sarah Palin and the Tea Party but then clutched their pearls at Donald Trump. Where the hell were their angry voices over the past hundred years as their party morphed into a modern-day Klan championed by racists, cheered on by idiots, normalized by the media, celebrated by foreign adversaries, and taken over by conspiracy theorists? Perhaps too busy trying to keep people who don't look like them out of the process. Isn't it ironic? Without us, the country they claim to love so much falls.

And the worst of this bunch are the people who rode the Trump train right to the nation's capital. But then when the monster inevitably turned on them, they stumped away expecting a welcome mat on this side of the divide. I'm talking about those who said nothing when he kicked off his campaign with a racist rant against Mexicans, the many who stood beside him when he called Nazis "good people," the heartless who call themselves "pro-life" Christians but defend the monster's policies against migrant children ripped from their parents' arms, sleeping on cold concrete floors with no proper bedding or nutrition. Do all lives matter or nah? When televangelist Paula White-Cain is telling them it's okay to love Donald Trump and his policies, did they ever stop to notice that the devil wears Prada? Or that the hate in which they cloak their "Christian conservatism" will eventually eat them alive? How sad that they only break faith when the monster attacks them. They're greeted with open arms by the media, who think nothing of their about-face and treat them like some sort of bigot-whisperer who can dissect Trump and his irrational behavior. Those people are not my allies; those people are my enemies. They can get the hell out of here with that "now we finally see Trump is no good" bullshit. Excuse us Black voters if we don't trip over ourselves to express gratitude to those who have abandoned the forty-fifth president but not his policies. Some of us are a little busy trying to bypass a white media landscape that largely ignores us and save the democracy the GOP is helping to destroy.

11

Silencing Black Women

Honestly, I can't always get with the feminist movement at large. Both the movement and the media's coverage of it seem overwhelmingly partial to the suffering of white women. In fact, for most of my life the face of the movement *was* white women. Gloria Steinem was the face of feminism, but not Shirley Chisholm. And I was an adult by the time Angela Davis was regarded as a feminist. One of the most transformative movements of our time, #MeToo, has pretty much been told through the voices of white women. Alyssa Milano and Rose McGowan were heralded as the faces of #MeToo and not the movement's founder, Tarana Burke. It was clear whose pain was initially being heard. Remember when Academy Award–winning actress Lupita Nyong'o put forth groundbreaking allegations against infamous Hollywood producer Harvey Weinstein? Probably not, because it got about as much attention as Mr. Knowles would if Beyoncé were anywhere in the room. "Harvey led me into a bedroom—his bedroom—and announced that he wanted to give me a massage," Nyong'o wrote in a *New York Times* op-ed where she detailed repeated unwanted advances and predatory behavior. "For the first time since I met him, I felt unsafe." This entire testimony was pretty much brushed aside by the media for more "sympathetic victims." Feminism has never felt inclusive. It seems to be, most times, a movement designed by and for upper-middle-class white women.

I was speaking at a conference in Philadelphia about the lack of diversity among decision makers in media. A white woman approached me after I got off stage to tell me that progress takes time. She said that from her perspective, there is indeed a lot of diversity in media. I asked her if she could name five women who hosted shows on cable news. She did, easily. Then I asked her to name five women of color who hosted shows on cable news. She could only name one. That's not to say that there aren't more. But this woman couldn't name more than one because there are so few. She went on to say her bigger point was that we cannot bring people along before they're ready. She suggested that "we unite as women, all women—no matter what color—against these men who are trying to control our bodies." But diversity . . . that takes time, she said. Girl, bye. That's what I said.

This is so often how today's feminist movement strikes me. Some want us to be each other's allies. But not on issues unique to African Americans—for those, us Black folk must continue to exercise patience. How many white women ever bothered to ally with the Mothers of the Movement—Black moms who had lost sons to gun violence, mostly at the hands of police.

How many have spoken out about pay discrepancies between white women and women of color? I've personally never been asked by a white woman *how* she could ally with *me*. I've only been given instruction on how I *should* ally with them. But I can't really sing in their chorus of grievances while, for the most part, they summarily dismiss mine. And ain't I a woman?

When Sojourner Truth posed this question more than a century ago, she was not met with welcoming arms from her white counterparts. Far from it. White women wanted her to advocate for *their* freedom to enter the ballot box as women but not for the general liberty of all Black people. In fact, suffragist Elizabeth Cady Stanton, the abolitionist daughter of a slave owner, acquiesced to the white supremacy on which she had been bred. She was a vocal adversary of the Fifteenth Amendment to the Constitution, which purportedly barred the states from denying Black men the right to vote. "We educated, virtuous white women are more worthy of the vote," Stanton declared at the time, according to historian Lori Ginzberg. "What will we and our daughters suffer if these degraded black men are allowed to have the rights that would make them worse than our Saxon fathers?" Stanton reportedly asked. Frederick Douglass, whose 1895 death stunned the Trump administration in 2017, retorted with a word. And a mood. "When [white] women, because they are women are hunted down through the cities of New York and New Orleans; when they are dragged from their houses and hung upon lampposts; when their children are torn from their arms and their brains dashed out upon the pavement; when they are objects of insult and outrage at every turn; when they are in danger of having their homes burnt down over their heads;

when their children are not allowed to enter schools; then they will have an urgency to obtain the ballot equal to our own."

Every year, white hosts, reporters, and commentators commemorate 1920 by saying it's when women got the right to vote via the Nineteenth Amendment to the Constitution. Not true. That year is when *white* women got the right to vote. Though women of color were technically included, still most were unable to vote. In the south—where 80 percent of Black women in the United States lived in 1920—they were still subjected to violent voter suppression. Black women had to continue their fight to secure voting privileges, for both men and women. But that caveat is rarely offered. It's as though our story never occurred—this erasure was by design. Black women were always relegated to the back of the line—literally and figuratively. In 1913, white organizers of the huge suffragist parade in the nation's capital demanded that Black participants march in an all-Black assembly at the back of the parade instead of marching with their state delegations. Consequently, many of the Black women who participated are intentionally almost entirely absent from photographs. Black reformers like Mary Church Terrell, Frances Ellen Watkins Harper, Josephine St. Pierre Ruffin, Charlotte Forten Grimke, Nannie Helen Burroughs, Ida B. Wells, Dr. Dorothy Height, and Mary McLeod Bethune are rarely recognized for their work. Susan B. Anthony got a coin. We can't get Harriet Tubman to appear opposite slave owner Andrew Jackson on the $20 bill. As Black women are paid 21 percent less than white women, not only can't we get that paper in our pockets, we can't even get our countenance on our cash.

There has long been a concerted effort to silence Black

women, whose positions on the frontlines of democracy have been transformative. Such was the case with Fannie Lou Hamer. She was tormented her entire life by white supremacy. And because of that, she had a testimony to share. But the white guardians of influence had their power to protect and could not risk her words disrupting their well-protected tenets of authority. Since they could control what hit the airwaves, they tried to ensure that the country would never know her story. A daughter of Mississippi, Hamer bore no biological children due to a hysterectomy performed by a white doctor without her consent while she was undergoing surgery to remove a uterine tumor—a common practice at the time, as an effort by white people to reduce the Black population. In fact, it was so widespread in the region that it was dubbed a "Mississippi appendectomy." She became an organizer for the Student Non-Violent Coordinating Committee (SNCC) and was a longtime voting rights advocate and organizer. She was fired from her job as a sharecropper on a plantation after she attempted to vote, and the plantation owner confiscated much of the property she shared with her husband. And this was only part of the story that white politicians never wanted America to hear.

Hamer was brutally beaten and arrested for attempting to eat at a white restaurant. While in jail, she and several of the women were brutalized again. "They beat me till my body was hard, till I couldn't bend my fingers or get up when they told me to. That's how I got this blood clot in my left eye—the sight's nearly gone now. And my kidney was injured from the blows they gave me in the back," she recounted. Not only was she beaten by a police officer while in jail, but the other officers ordered two Black prisoners to beat her as well. At the exact time she was being brutalized in a Mississippi prison cell, just up the road white supremacists shot and killed civil rights activist Medgar Evers.

By 1964, Hamer cofounded the Mississippi Freedom Democratic Party (MFDP), which challenged the local Democratic Party's efforts to block Black participation and demanded to be recognized as the official delegation. Hamer spoke before the Credentials Committee, calling for mandatory integrated state delegations. But President Lyndon Johnson, who would owe his first elected term as president to Black voters, was terrified that she would alienate the support of southern Democrats, which he also needed to win reelection. "I do not believe I can physically and mentally carry the responsibilities of the world, and the Niggras, and the south," he said to White House advisers. He sent some of those same advisers to appeal to her to be silent. She refused. So he silenced her instead. He intentionally held an impromptu televised press conference to preempt Hamer's remarks, ensuring she did not get airtime. His trick worked—but only briefly. Her words addressing voter suppression and state-sanctioned violence were so powerful, they were broadcast anyway later that night. She detailed the agony she had endured. In conclusion, she asked tearfully, "Is this America? The land of the free and the home of the brave, where we have to sleep with our telephones off of the hooks because our lives be threatened daily because we want to live as decent human beings, in America?" Hamer's testimony, which aired during the prime-time news, would become one of the most powerful speeches of the civil rights movement. By 1968, she was a member of Mississippi's first integrated delegation.

Today, there is an uneasy pathway to giving voice to Black women and shaping how our stories are told. Women of color represent just 7.95 percent of US print newsroom staff, 12.6 percent of local TV news staff, and 6.2 percent of local radio staff. Cable news and network outlets don't publicly release the racial breakdown of their employees. However, these networks have consistently shown where their preferences lie both in coverage and employees—and both seem to be painfully out of step with the changing demographics of the country. Two months after Jeff Zucker took over CNN, he quickly removed Soledad O'Brien from the morning news anchor chair, replacing her with a younger white woman. Over at NBC, as mentioned earlier in this book, executives favored the blackface apologist Megyn Kelly over Tamron Hall—the first Black woman to host the *Today Show*. Both O'Brien and Hall were offered smaller roles within the company, and both declined. In June 2016, the *LA Times* ran a story "The Women of MSNBC Are Reshaping the Television Landscape," featuring Stephanie Ruhle, Katy Tur, Hallie Jackson, Nicolle Wallace, and Andrea Mitchell. No one seemed to notice that while women of color make up nearly 40 percent of all women in America (and Black women comprise a sizable portion of MSNBC's viewing audience); none were featured in this story. Did no one at the *LA Times* notice the lack of diversity? Did anyone on the PR team at the network notice? Did the talent themselves notice? It's not that these women didn't deserve to be celebrated—but what exactly was the story? Name me any cable news network that doesn't have five white women anchoring programs.

Women of color simply do not get the same opportunities

or treatment, and as a result, Black women and our perspectives are left out. White managers tend to hire other white people because they're more familiar with them and their work. Their social and professional network is insular, and we are rarely an intimate part of it. This is part of the reason why social media outlets have become such a credo in the Black community. At times, this is the only place you can get Black perspectives. When President Barack Obama came into office, there were slight ephemeral bumps in diverse on-air talent. But when he left office, the diversity efforts seemed to fade to black. The Black women who remain as on-air talent, executive producers (of which there are very few), producers, bookers, and writers are rarely given editorial decision-making power. And this has certainly impacted the media's coverage of an increasingly diverse political landscape.

Mainstream media outlets had no choice but to acknowledge the Black women in Alabama who stopped an accused pedophile from making his way to Capitol Hill and instead helped the state send a Democrat to Congress for the first time in twenty-five years. Democrat Doug Jones's win over Republican Roy Moore was only possible thanks to the 98 percent of Black women and 93 percent of Black men backing Jones. But were they so enamored with the white southern Democrat? No—and that's what was missing from the reporting on this story. The alternative was a Trump sycophant who had faced allegations from six women who accused Moore of pursuing romantic or sexual relationships with them when they were teenagers as young as fourteen and he was an assistant district attorney in his thirties. Two of the girls accused him of assault or molestation. Moore denied the allegations, dismissing them as "fake news." And he enjoyed the support of the nearly two-thirds of white women in Alabama who cast votes for him. The Black women in the state were busy also electing nine Black women as judges that year, highlighting their power to shift local and state races as well.

During the 2018 midterms, Black women voted overwhelmingly for Democrats (90 percent) and increased their own share in Congress from 4 to 5 percent, helping Democrats regain control of the House. The US Census Bureau reports that 55 percent of eligible Black women voters cast ballots in November 2018, a full six percentage points above the national turnout. Black women—both as candidates and as voters—reigned supreme. A report from the Reflective Democracy Campaign found that in the race for Congress in 2018, there was a 105 percent increase in candidates who were women of color. They made up 4 percent of candidates and 5 percent of the winners. The 116th Congress was the most diverse the country had ever seen. Meanwhile, white men comprise 30 percent of the population but still hold 62 percent of elected offices at the local, state, and federal levels. The political structure still overwhelmingly favors white men as the ruling class. But this is frequently overlooked or ignored by many in the media landscape, because the uneven power structure so closely mirrors their own.

A White Woman Narrative

The master narrators who have editorial decision-making power found a way to paint Melania Trump, a complicit defamer, as a pushover patsy. The former nude model who, before she became First Lady, appears to have been the second lady when she met the reality TV star, as he was still technically married at the time to Marla Maples. Melania became a US citizen under suspicious circumstances: she obtained an "Einstein visa," reserved for immigrants with "extraordinary ability" who have "sustained national and international acclaim." She was just one of five people to get approved for such an honor in 2001—just a few years after she began dating Donald Trump. I've seen the nude photos—they're googleable. And I ask, what do we think her special skill was that afforded her the ability to leapfrog over actual geniuses to secure American citizenship? Becoming a citizen allowed her the right to sponsor her parents, who are now in the United States and in the process of applying for citizenship. You know? That chain migration that the Trump administration consistently railed against? I'm sure the anti-immigrationists who populate the Trump administration and his rallies are just downright outraged.

Melania perpetuated with glee and enthusiasm the racist theory that Barack Obama was born in Kenya. She plagiarized a Michelle Obama speech at the 2016 Republican National Convention. As First Lady, when visiting migrant children she wore a jacket that read, "I really don't care do you?" The entire notion of hypocrisy seemed painfully lost on her when she launched her "Be Best" campaign aimed to reduce online bullying. (So we're just going to pretend that "Be Best" is proper grammar? Okay.) The CNN reporter who covered FLO-TITS rarely made mention of any of these things. She was presented to the world by the American press corps as someone who was constrained by her husband's irrational thoughts and not as someone who shared them. Cable news outlets posted online articles like "Melania Trump: From Soft-Spoken Schoolgirl to Next First Lady," and certain reporters who cover her wrote books reimagining her as someone who needs to be freed and not someone who is enjoying a life of her own making. Not someone who excuses her husband's brash insults and echoes her own. Where the media saw a victim, we saw a villain.

Meanwhile, during her tenure as First Lady, Michelle Obama was criticized for everything from her fashion choices (Gasp! She attended the State of the Union in a sleeveless dress!) to her White House initiative as First Lady (Childhood obesity? Let them eat cake!). Many in the conservative media disparaged her physical appearance and dredged up deep-seated stereotypes of Black women that are not worth taking up space here.

During the 2008 presidential election, I remember hearing countless discussions about Hillary Clinton and Barack Obama. The question of who was being treated more fairly on the campaign trail essentially gave way to the subtext of who is treated more fairly in general: white women or Black men? This preposterous question didn't warrant a debate. There is certainly intersectionality in our politics. But where racism and sexism intersect, I am often rebuffed by members of my own gender tribe when it comes to my race. Who has it worse? Black men or white women?? I won't legitimize such a ridiculous debate by offering innumerable examples throughout history that should immediately dismiss the discussion. Let's not do comparative suffering. But guys . . . come on. Stop it with that.

In 2017, the Women's March was criticized for being over-whelmingly white, despite a leadership group that was primarily women of color. Observers initially denounced the march for failing to pay enough attention to issues of race. Still, some of the white women said they felt alienated. The *New York Times* published a story about some white women feeling excluded by demands that prioritized awareness of racial and economic inequality. They feared that addressing issues of race would drive away potential participants. When a post on the march's Facebook page quoted Black feminist Bell Hooks about forging a stronger sisterhood by "confronting the ways women—through sex, class and race—dominated and exploited other women," one of the participants replied, "I'm starting to feel not very welcome in this endeavor." Women's March Wendy likely doesn't feel welcome anywhere she's not centered.

And that's really the problem here, isn't it? Some white women wanting to feel centered and most white men's exclusive promotion, celebration, and reinvention of them. This is the same group who so frequently advocate for equality and then actively engage in or benefit from oppression.

12

Kamala's Campaign

Okay, can we first admit that as a Black woman running for president Senator Kamala Harris would have had to essentially be a supreme being with an unflawed background and run a near perfect campaign to actually clench the 2020 Democratic nomination? Harris was interviewing for a job that had previously been held by all white men, except for one. Obviously she was not the only Black candidate in the 2020 race. She was joined by fellow senator Cory Booker of New Jersey and mayor of Miramar, Florida, Wayne Messam—both of whom eventually dropped out of the race. However, Harris's candidacy captured well some of the fractured perspectives in the Black community and highlighted distinctions between how mainstream media covers women candidates of color and those whom they deem viable.

Though the senator is remarkably impressive (and an absolute delight in person), I will admit, Harris did not run the best campaign, so this is certainly not to suggest that had all things been equal she would have had smooth sailing to secure the democratic nomination. Not by any means. And if you're a Harris supporter, do know that I obviously support elevating and empowering Black women. However, just because I am a Black woman does not mean I am obligated to perpetually support the candidate of your choice while never saying anything critical (that's not for everyone reading this—it's mostly for the Twitter crowd).

When Harris announced her campaign, I thought the coverage was pretty favorable—in fact, too favorable. She was very savvy to string out the announcement of her candidacy over a week, which included a press conference at her alma mater, Howard University, and culminated with a crowd of over twenty thousand people in Oakland, California. Then came the reporters making the obligatory comparison to Obama—an expectation, impossible to meet, that was set by the media. But soon the enthusiasm faded. The 2020 field all had to respond to concerns about electability through the lens of being able to beat Donald Trump, but Harris had to field assumptions about electability with the added punctuation of being a Black woman.

The media frequently gets Black voter loyalty wrong in their diagnosis of our political preferences. It's not that Black voters overwhelmingly like or dislike one candidate or another. For many, the power of white supremacy is so prevalent in the functioning of our democracy that Black voters never want to chance their vote on someone white America will not support. And this is done in the *Democratic* primary, emphasizing the point that Black voters have always known that racism is bipartisan.

Harris understood that her campaign would not advance

without the support of the Black community. Appropriately, she appeared very proud of her Black experience. But ironically, many parts of *her* Black experience alienated her from some sects within the Black community. Her pedigreed background was certainly an authentic Black existence. As a graduate of Howard University, an HBCU in Washington, DC, affectionately dubbed the Black mecca, surely she was in the diaspora. It was there that she pledged Alpha Kappa Alpha sorority, the first Greek-letter organization for Black women. She went on to attend Hastings College of the Law at the University of California. There are innumerable Black people who find this path very familiar, of course. Black voters who summer on Martha's Vineyard, brunch at fancy restaurants that are not chains, shop at Nieman's and Saks department stores, and have a few designer shoes in their possession—they completely *get* Harris's background. Many of them share a similar path. But this is not everyone's Black experience. In fact, many found it completely unrelatable. In some Black homes, talking about your private college and Black sorority experience is not going to resonate. Voters in this bloc may not max out to candidates in campaign contributions, but their support is just as integral.

Then there was the question of Harris's ethnicity. Was she Black enough? This was a question that some in the Black community had unfairly posed earlier. Just like President Obama, Harris is of mixed race. Her mother was South Asian, from India. Her father is Jamaican. Both her parents were academics who were active in the civil rights movement as graduate students at the University of California, Berkeley. But for some Black people, when our brethren's ethnic makeup and upbringing don't mirror their own, it garners an eyebrow raise. Upon seeing Harris, my assumption would be that I am looking at another Black woman. But some found her racially ambiguous appearance questionable and her general disposition inauthentic. Almost immediately after she announced her candidacy, many on social media elevated these questions, casting doubt on her authentic Blackness. In fact, Harris generated significantly more attention on Twitter than others in the field, according to research compiled in the report *#ShePersisted: Women, Politics and Power in the New Media World.* The dominant narrative focused on her identity, attacking her as "not authentically American," because both her parents were immigrants to the United States, as well as not authentically Black. The bulk of these tweets originated from trolls and fake news accounts before penetrating the atmosphere.

Harris arguably had done little to engage the Black community leading up to her presidential run. It seemed as though she assumed Black America knew her, and many did not. It's a common mistake to be beckoned by the glory of mainstream white-run media and forgo smaller niche markets that speak to sects of the Black community—the very community she needed to elevate her prospects. She reserved that level of engagement for the campaign trail, despite having some meme-worthy moments in the Senate. It would have made sense to elevate those in front of a receptive audience and shore up support.

And when she *was* covered in the media, it was a learning curve for white reporters. In January 2019, *Washington Post* journalist Chelsea Janes was covering an event for Harris's memoir and tweeted about "members of her Howard sorority" (AKAs) who began "screeching" (skee-wee, the AKA call) and adding, "I didn't expect to hear that here." This is not Janes's fault, per se, but it certainly highlighted how reporters of color must be familiar with every part of a society that is not inclusive of our culture, while white reporters have the privilege of being excused for their ignorance about many parts of our culture. Like when *Politico's* Michael Kruse tweeted video of then 2020 contender Andrew Yang dancing with a group of women in South Carolina, describing it as "jazzercise" when it was actually the "the Cupid Shuffle."

During the early stages of the campaign Harris appeared ill prepared to offer a solid vision on healthcare. She was vague when responding to questions about her plan, and for this she was hammered in the press. Nothing really unfair about that. However, when fellow Democratic contender Senator Elizabeth Warren spent months being vague about her healthcare ideas, she was not met with the same level of scrutiny. There was also an assumption in the media that Black voters would automatically support one of the two Black candidates in the race. The question "Why aren't you doing better with Black voters?" was almost exclusively reserved for Harris until they decided to dismiss her campaign and focus on former 2020 candidate Pete Buttigieg. Many candidates in the field had yet to register with a large bloc of Black voters. Again, this punctuates the point that there should be no umbrella for Black voters nor should there be assumptions.

Then there was the media's love affair with the trio of white male candidates. The landscape swooned over Buttigieg, whose only political experience was being mayor of the fourth largest city in Indiana, a city whose entire population could fit in a college football stadium—and of course running and failing in statewide and DNC leadership contests. The media showered him with praise before he actually earned it. There were echoes of Biden's electability and his magic formula for bringing Trump voters to the Democratic side of the divide. And though he did eventually rise to become the Democratic nominee, he too was aided in favorable media coverage as many saw him as more electable than a woman or a person of color. And finally, there was Vermont senator Bernie Sanders, whose grumpy persona was often viewed as a lovable grandpa while every other week there were "likability" conversations about the women in the race. There were so many stories, in print and broadcast, that simply left Harris out entirely. The stories that did focus on Harris discounted her campaign altogether. Given the media's undue influence on voters and their perceptions about candidates, their critiques and predictions can become self-fulfilling prophecies.

There's a popular theory among some longtime political strategists that the responsibility to connect with voters always lies with the candidate. So if Harris didn't connect with Black voters, she's entirely to blame? I beg to differ. I think there's some accountability for everyone. When we talk about the media's treatment of the only Black woman in the race, we also have to look at what some Black voters chose to hear. With an influx of things constantly begging for our attention, the instinct is to look for the most salacious thing. And that same instinct proves to be true in our politics. Cory Booker introducing a policy to ban facial recognition technology from public housing isn't nearly as interesting as a photo of him and Rosario

Dawson showing up hand in hand at a fundraiser. But since Black voters are likely more impacted by policy, we have to be more disciplined in our focus.

Such was the case when Harris appeared on the syndicated radio show *The Breakfast Club*, a popular morning show that became a must-stop on the campaign trail during the 2020 election cycle. During an appearance in February 2019, Harris said on the show that she had smoked marijuana in college. "What were you listening to when you was high?" asked host Charlamagne Tha God. "What was on? What song was on?" DJ Envy weighed in and asked, "Was it Snoop?" Harris replied, "Yeah, definitely Snoop. Tupac for sure." A number of people cried foul, pointing out that Harris graduated from Howard University in 1986 and finished law school in 1989, but Snoop Dogg and Tupac's first albums didn't come out until the early 1990s. Well! By all means! Somebody call Perry Mason. We've got a case to crack.

Guys, if you asked me to recount what I was listening to when I was smoking weed a year ago, I wouldn't remember. Asking me what I was listening to decades ago? That's nearly impossible. If your thoughts were that clear, you might need to find a better weed guy. But more importantly, during this same interview, Harris also discussed the LIFT Act—legislation she introduced in Congress that would provide up to $6,000 a year in the form of a refundable tax credit to middle-class and working families. She detailed during the interview how the legislation would lift 60 percent of Black households out of poverty. Not a lot of discussion about that surfaced. Look, I get the instinct for some younger people to pounce on something like misrepresenting what music one was listening to when blazing. Fine. Far too often, politicians parade themselves before Black audiences and, like chameleons, try to inauthentically blend in and say only what they think we want to hear. But in our discernment during those times, I'd argue that same critical thinking should overtake some people's desire to look only at the Twitter-worthy, parsed down sound bites focused on pop culture. We've got to be a lot less interested in what Harris was listening to when she sparked up a J in the 1980s and much more interested in her outlook on marijuana legislation.

Which brings us to her time in California as a prosecutor, when some of her policies negatively impacted people of color. But before we get into the specifics, it's important to point out that we're viewing these policies through a lens decades removed from what was happening at the time. The newly woke coalition who are perhaps too young to recall the turmoil of the 1990s or have simply forgotten should revisit the era. There were Black mothers and grandmothers begging for help to make their communities more safe. These were not policy experts charged with foreseeing the unintended consequences of laws. They were matriarchs pleading for a better day so that their children and grandchildren might survive. They were asking a system designed to harm them to instead protect them. Similarly, Harris argues that she was trying to appease these requests while also trying to disrupt that same system from the inside. She never apologized for her record, perhaps because she felt she had nothing to apologize for. On the campaign trail, Harris recalled spending time with grieving mothers who wouldn't talk to anyone in the DA's office but her. She defended her legacy as a prosecutor, defiantly saying she was combating injustice and racial bias. This is neither a defense nor a criticism of her record. Just context.

Harris had positioned herself as a "progressive prosecutor." As district attorney, she pledged to never impose the death penalty, but she once fought to overturn a ruling that could have helped to wipe out the death penalty statewide—a ruling she said was not compliant with state law. Harris wrote a book called *Smart on Crime* that urged officials to abandon policies of the past and instead favor rehabilitation over punishment. But some argued her positions were only slightly less awful than the ones she was looking to change.

Law professor Lara Bazelon penned a scathing opinion piece in the *New York Times* that was published days before Harris formally entered the race. It arguably lacked perspective, context, and research but was shared widely. Under the headline "Kamala Harris Was Not a 'Progressive Prosecutor,'" Bazelon writes:

> Time after time, when progressives urged her to embrace criminal justice reforms as a district attorney and then the state's attorney general, Ms. Harris opposed them or stayed silent. Most troubling, Ms. Harris fought tooth and nail to uphold wrongful convictions that had been secured through official misconduct that included evidence tampering, false testimony and the suppression of crucial information by prosecutors.

Harris was also widely criticized for her handling of San Francisco's rampant truancy policy, which statistics showed when policy wasn't enforced it led to an increase in the dropout rate and victimization. As district attorney, Harris responded by threaten-

ing criminal charges against parents of truant students, though
she says she never actually sent any parents to jail. Despite concerns that the policy would disproportionately affect low-income
people of color, she pushed the policy statewide as attorney general of California, and that's when some jurisdictions did send
parents to jail while enforcing the policy. "I refused to stand by
while the system failed them [the students]," Harris said in a
2019 speech in South Carolina defending her record. "So I held
the system accountable and got those kids back in school—not
by sending people to jail, but by getting families the resources
they needed. Those children deserved justice."

Harris supported the use of medical marijuana but refused
to support recreational use—that is, until 2018. Critics also say
Harris squandered the opportunity to improve police accountability statewide by deferring to local prosecutors regarding the
use of police body cameras and investigation of police shootings. Although she did support the use of body cameras, she
opposed a bill that would have required the attorney general's
office to investigate police shootings. "I made the state law enforcement agency under my supervision wear body cameras to
increase accountability to the community," Harris said, trying
to explain her stance. "I launched investigations into acts of discrimination by law enforcement agencies. All because the people deserved justice." She did not address why she opposed the
bill. All of these questions on her positions perpetuated perceptions that she lacked an ideological core on policy and instead
chased public opinion.

There was not a lot of space to unpack her true stance on
these policies when it's a crowded field of presidential contend-

ers clawing for attention. Far too often, a candidate's entire political history can be trivialized by an unflattering headline or a ninety-second news package. So some Black voters simply diluted Harris's story to "she locked up Black parents, sided with police officers, and used this to advance her political career." Fair or unfair, accurate or inaccurate, the media spent little time informing voters on these issues and by the time these headlines ran, voters were not enthusiastic about Harris as president.

The ways in which Harris was able to influence policy that did positively impact people of color did not get as much attention. She pioneered a plan offering job training to incarcerated first-time offenders. This plan has since been adopted around the country as a tool to combat recidivism. As San Francisco's district attorney, she challenged California's controversial now-overturned Three Strikes law, which allowed sentences of twenty-five years to life for a third felony conviction, saying she would enforce the sentences only if the third offense was serious or violent. As the state's attorney general, she mandated implicit bias training. She was also commended for her work in correcting a backlog in the testing of rape kits.

In a speech addressing her record as a prosecutor, she tried to make clear that she understands the concerns around her career choice. "There have been prosecutors that refused to seat Black jurors, refused to prosecute lynchings, disproportionately condemned young Black men to death row, and looked the other way in the face of police brutality," she said. "And I was clear-eyed that prosecutors were largely not people who looked like me . . . or the people who grew up in our neighborhood." She was only one of three Black district attorneys in the entire office in San Francisco. "It matters who is in those rooms," Harris said. "I knew I had to be in those rooms. We have to be in those rooms when there aren't many like us there." She highlighted cases in which she prosecuted child molesters, rapists, murderers, banks that defrauded homeowners, polluting oil companies, and pharmaceutical companies. But for many Black voters, this counternarrative was not enough.

There's a hierarchy of accountability that leads some voters directly to Harris. Though some were able to overlook policy choices from candidates like Joe Biden in the 1994 Crime Bill and the treatment of Anita Hill, they could not look past what they considered to be Harris's shortcomings. I surmise that some expect more from members of our own community. There's a willingness to forgive white people for doing something racist in their past, in the same way that you might forgive a puppy for peeing in the house—neither yet knew any better (though I beg to differ). Newly woke white people were welcome to our votes, but a Black person who made a policy misstep should have foreseen the unintended consequences that would harm her brethren. It's a subconscious philosophy that the system designed for our demise can never truly be beaten employing traditional methods. So any attempt to disrupt it from the inside is therefore moot. This by no means represents the outlook of all Black voters, but it was certainly a lens through which a tiny few viewed Harris.

After all, a fellow 2020 contender, Minnesota senator Amy Klobuchar, also had a sketchy past as a prosecutor in heavily white Minnesota, where she consistently declined to go after police involved in fatal encounters with Black men. Where are the scathing op-eds on her? To be fair, there may be none because most Black voters didn't know who she was—something Klobuchar didn't seem to aggressively want to address. But back to Harris . . .

There was the complicated relationship she had with Black men. One of her more vocal critics was 1980s rapper Luther "Luke" Campbell. He ultimately became a surrogate for Harris, but before that, he had penned a harsh op-ed in a Florida newspaper criticizing Harris's record as a prosecutor and unfairly attacking some of her dating choices—another distinction unique to a Black woman candidate. Like many Black women, Harris has been single most of her life, which meant she had a dating history. She briefly had an affair with San Francisco mayor Willie Brown. But context is important here. Brown and his wife had been separated for years. This was not a tawdry affair built on clandestine hookups at seedy hotels. If you were picturing Olivia Pope–level passion with a powerful man, you will be disappointed. Their relationship was very public. Luke's op-ed didn't mention that part.

Nonetheless, I referenced the op-ed on MSNBC's *AM Joy*. I said this might be a problem for her and suggested that she consider adding a Black man as a validator on her team. My thinking was someone who could offer a different perspective to the many Black men who were not willing to offer their support to Harris at the time. I was quickly criticized for lending legitimacy to Luke, whom many believed to be misogynistic. Publicly and privately, I challenged these very vocal critics. I was not endorsing Luke's point of view. However, last I checked Florida was a battleground state. Luke runs a football league that reaches thousands of parents, writes a column that reaches thousands of readers, and has influence with tens of thousands of people—mostly Black men—across the country. Were I working on a campaign, I would not get in the habit of dismissing or silencing such a voice.

But that's not to say there is no sexism within the Black community. Far too often when we think of sexism, the accompanying image is white men. But some Black men definitely have gender bias issues as well. Take the gubernatorial race in Georgia, for example. According to exit polling provided by CNN, 11 percent of Black men voted for the Republican candidate and well-documented practitioner of voter suppression Brian Kemp. And in 2016, Donald Trump earned double-digit support from Black men in the presidential election. It's not a popular thing to say, but Harris *did* have a Black man problem. But we must also examine that some Black men have a Black woman problem.

There was the issue of Harris's marriage to fellow attorney Doug Emhoff, a white man—a major point of contention for many Black men. So many posed the question to me, How is it that Kamala Harris went to an HBCU like Howard University and still chose to marry a white man? Okay. There's a lot to unpack here. One, the question itself puts the onus on Harris—and by extension any Black woman who marries outside of her race—and suggests that marrying anyone other than a Black man is somehow a character flaw, as though it's a personal failing. The question was *never* how did Black men let Kamala Harris get away? Moreover, statistically, Black men marry outside their race at significantly higher rates than do Black women. Yet this source of outrage was almost exclusively reserved for Harris.

When asked about her marital choice, Senator Harris gave the only reasonable response she could: she married a man who loves her. Are we seriously to blame Black women for finding love? If it's in the form of a white man? This is in no way to say that Black men don't love Black women. But it seems an unfair thing to consider when weighing the next leader of the country. After all, this was a presidential election, not *The Bachelorette*.

The plain truth is that Kamala Harris is an accomplished politician with a flawed past who ran an imperfect campaign. That makes her no different from any of the men who have been elected president. If a racist half-witted liar who proved to be an existential threat to America could find his way to the White House, why not a capable Black woman? If the choice were between Harris and Trump, I firmly believe a coalition would have gathered around this historic candidate to help usher her across the finish line. She would have outsmarted Trump on the debate stage and embarrassed him with intellectual clapbacks—because so often, as Black women, we have to think on our feet, keep our cool, and tell someone to fuck off with smiles on our faces.

There were people who wanted to like Kamala. More importantly, they wanted to vote for her. But they needed a reason beyond policy. They not only needed to believe her. They needed to believe *in* her. For an election in which a Democratic challenger would face Donald Trump, voters—Black voters in particular—wanted a sure thing. For that person to be Harris, she would have been required to transform the myth of "Black girl magic" (which is really years of hard work after a lifetime of having it drilled into our heads that no matter how hard we work, we will never be considered good enough, so we become too great to be ignored) into superhuman abilities. After making concessions for a past of which she seems proud, she would then have to penetrate the fear among Black voters to convince them that white people would indeed support her—a Black woman.

We can parse over her history, both personal and political. We can debate every single mistake she made on the campaign trail. We can comb through her legislation on Capitol Hill and pick apart her policy proposals. But the media—and voters— tends to set such a low bar for the Republican champion they send before the people to try and best the Democratic oppo- nent, for whom there's such a high bar. The bar gets higher when it comes to Black candidates and becomes a nearly insur- mountable hurdle when it comes to Black women.

As Harris's poll numbers declined, the media expressed dis- appointment compared to their hyped-up expectations. Then they paid her considerably less attention, save for those hit pieces that were running pretty regularly. Then her poll num- bers plummeted more, because the media was giving her less attention. As a result, voters were hearing less of her. This made it harder to fundraise from a donor class that's largely white and male. All of this created a path of hurdles she found insur- mountable. Eventually, she ended her campaign. By the time the dust settled, there was not a single person of color left on the path to the White House. Interestingly, by the time the field narrowed to all white men, there was a strange nostalgia for her and regret that she was no longer on the stage. Perhaps having the benefit of hindsight, voters would find her more palatable as vice president.

THREE

Where We Need to Go

13

If You Don't Vote, You Don't Count?

I f you don't vote, you don't count. Can you imagine a more disrespectful phrase to utter to a Black person in America? As though our birthright to exist in this country can be reduced to a vote, a system that was never designed to include our participation. I admit, I used to believe these words myself. But as I matured, I came to better understand what Black American patriotism looks like. It's a complicated sentiment interwoven with pride, sadness, battle scars, determination, heartache, community, fury, love, and righteous anger. Moreover, I've laid out numerous scenarios in this book to prove that even when you do vote, sometimes you don't count. It's a process to convince people who have been systematically shut out for so long to prioritize a democracy that for so long did not include them. "You're telling people to believe in something that mistreats them, that has exploited them and their family, that has locked up their brothers. And now we're asking them to put their faith in that same system that they're looking at every single day that keeps them at a disadvantage. That sounds dishonest," LaTosha Brown says. "We have to convince Black people who have been so disenfranchised to imagine a new America." Given the election of Donald Trump, which highlighted the flaws in this democracy, it will be challenging for the country to continue to exist successfully in this framework. So what does the next phase of democracy look like, and what will be the role of Black people in its shaping?

This is the question Brown and her organizers present to convince masses of Black people across the country to vote. "I am not trying to convince people to believe in this system that we are experiencing, which is structurally flawed. My objective I share with people is this: I need you to believe in you, brother. I need you to believe in you, sis. It's a mission to get people to believe that you have agency over your life because all that other shit told you that you don't." Brown says this is when she sees people shift, and then they are unstoppable because the outcome is not simply about their vote, and they have not simply been reduced to a part of a system. "The outcome is getting them to believe in themselves in a different kind of way. The outcome is what gave Black folks the valor after slavery to lead three hundred Black folk in Mississippi, though in a week's time seventy of them were killed by white mobs. What would make you do that? Because you felt a sense of your agency."

Black Voters Matter is building infrastructure that is not specific to one candidate or one political party, but designed to maximize the engagement of Black people in the community. The group has organized numerous communities across the country, providing mini-grants to encourage communities to vote. "Voting won't be the great Liberator," says Brown. "But we do think that voting is a tool for harm reduction. If any decisions are being made related to our community, we should have some influence and impact on those decisions." Getting Black people organized around that concept is self-preservation for Black people. Nothing highlights the need for harm reduction more than the failed war on drugs, which was essentially a war on Black people.

Consider the media's coverage of, and the political response to, the opioid epidemic. Contrast that with the crack epidemic of the 1980s. This discrepancy emphasizes the racialized deployment of the war on drugs, which needlessly ensnared millions of Black people in the criminal justice system and devastated communities and generations. Black people were not afforded a gentle, kind, sympathetic justice system, media landscape, or political system. In fact, President George H. W. Bush sought to punish everyone involved in any part of the drug trade to the fullest extent of the law. And he used the power of his budget to do it, with the sharpest increase in funding almost specifically designed to jail and kill Black people. When Bush took office, the federal drug control budget was around $5 billion. It was over $12 billion by the time he left in 1993. Those funds were used to double down on policing. He said the solution was "more prisons, more jails, more courts, more prosecutors." This war on Black people would fundamentally restructure political power through unprecedented incarceration rates and disenfranchisement laws. We're now asking Black people who were targets of this system to believe in it.

So when politicians speak about democracy as service to America, it falls flat in our community, where a debt is still owed. One of the most beloved presidents in the country coined a phrase when he directed, "Ask not what your country can do for you, but what you can do for your country." That's not everyone's belief. "You can ask that when you come from a place of privilege," Brown says. "When your needs are met and your family has wealth, which frankly other people created, you can have this charitable patriotism." But Brown argues that the fundamental belief in democracy is supposed to be for the people, of the people, and by the people. "So if that is the case, the explicit nature of government should be to serve its people. We have asked Black folk who have created the wealth in this country, who have created the very tools that the privileged benefit from, and Black folks are supposed to continue to give their labor, their blood, sweat, and tears and not ask their government for anything? How do you figure that?" How indeed.

Reaching Black Voters

Many candidates, particularly at the federal level, have a bad habit of talking *at* and *about* Black people. Far too few have actually engaged in any meaningful discourse. And since the media landscape continues to center political conversations around the needs and desires of white people, there's a gross lack of understanding about how to engage Black voters. The stale and dated efforts to appeal to the constituency have long included an obligatory stop on the campaign trail to break bread with the lovely Reverend Al Sharpton in Harlem, visit churches and sit with pastors in South Carolina, and meet with small groups of journalists who write for Black publications. Oh, and let's not forget, as of late, the mandatory pro-reparations declaration. Candidates have previously been soft on meaningful policy, and that shortsighted approach has left out millions of Black people across the country who may never read niche media outlets or any papers at all. As coverage of issues relevant to Black voters increases, so too does the focus and engagement of candidates. Also, if you aren't one of the people filling a pew every Sunday, you can feel overlooked. "When it comes to Black people, the first thing people think about is barbershops, beauty salons, and churches," says Nadia Garnett, a political strategist who ran Hillary Clinton's Black voter outreach.

"We have to start assessing where we are and think beyond. That could be engaging young professional events. Do some small business roundtables. Black women are the largest group of small business owners and entrepreneurs and have been for years. Who's talking to them? What does the policy we're proposing look like for them?" She's correct. Black women are the only racial or ethnic group with higher business ownership than their male peers, according to the Federal Reserve. There were 2.4 million African American women–owned businesses in 2018, most owned by women aged thirty-five to fifty-four. And according to a report commissioned by American Express, the number of Black women–owned businesses grew by a whopping 164 percent in 2018 alone.

Garnett went on to say that candidates should not shy away from speaking to organizations that are 501(c)(3)—nonprofits that, to maintain their tax exempt status, steer clear of political campaigning. She says these groups —like the Links, Inc., a Black women's organization of financially comfortable women focused on community improvement; the Boule, a Black men's organization that accepts members who have gained considerable respect within their chosen industries; or the Guardsmen, a Black organization for like-minded people founded by HBCU graduates—represent politically active voters. Grassroots outreach should extend to all aspects of community. "Grassroots does not only mean people who are struggling," adds Brown. "To me that means organic community leadership. There are groups organizing grassroots suburban white women, right, that doesn't necessarily mean poor white women. Grassroots organizing means people in the community who may or may not necessarily be engaged, starting literally from the neighborhood level and up."

Meet us where we are—and we are everywhere. Political candidates will spend nearly $6 billion dollars on advertising in the next election cycle. Advertising Analytics, a nonpartisan firm that tracks the ad industry, estimates that campaigns will spend almost twice as much on broadcast and cable TV ads as was spent during the 2016 presidential campaign cycle. Former New York City mayor Michael Bloomberg made headlines when he entered the 2020 presidential race by dropping more than $30 million for one week's worth of television ads across the country—which basically bought him a rotation of three ads in ninety-nine local markets and a national audience. But if you're focused on reaching Black voters, this investment was about as valuable as a luxury tent at rapper Ja Rule's fraudulent Fyre Festival. Many people are not watching broadcast television as much anymore—they're streaming. And African American consumers specifically stream content at a higher rate than all other consumers, according to a 2018 report from Horowitz Research. This study found that nearly three-quarters (74 percent) of Black TV viewers report streaming at least some of their TV content, compared to 68 percent of all television viewers. Additionally, 65 percent of African American streamers say they watch more TV content than they did five years ago. Not surprising, considering that a Nielsen Total Audience Report for the first quarter of 2018 reported that African American adults consume nearly thirteen and a half hours of media per day, almost two and a half hours more than the average adult. "Since we are now viewing shows on our devices, you have to tap into that space," says Garnett. It's a space where, for Black people, the content matters. African American viewers overwhelmingly tune into content that features their images and stories, particularly on digital platforms. So, asks Garnett, "what type of ads are we placing during *Greenleaf*? And as we know, a lot of our HBCU games aren't broadcast on traditional TV stations. Our games might be on ESPN-U. So what does that look like for the folks who are trying to watch the games that aren't on ABC, NBC, CBS?" Horowitz reported that 64 percent of Black streamers say they will check out a show if it features a Black character or cast, and 57 percent say that shows featuring racial or other sociopolitical themes appeal to them. This can also apply to the many presidential debates. Black viewers, according to this evidence, will likely tune in when they see someone who looks like them. If there are Black debate moderators asking questions relevant to the Black American experience, this is also inviting to the millions of African Americans helping to shape democracy. Unfortunately, very few Black faces are allowed to use their authentic voices when moderating such debates.

Another space that's overlooked is gaming, Garnett says. According to a study released by Electronic Entertainment Design and Research in 2018, 66 percent of the general US population described themselves as gamers. And Nielsen found that African Americans in 2015 made up the second-largest demographic in the genre (though only 1 percent of people hired in the gaming industry are Black, but I digress). "So how are we getting into some of these gaming applications and embedding ads there?" asked Garnett. This is a trend that is catching on quickly. Twitch, which Amazon purchased for nearly $1 billion ($979 million) in 2014, is primarily a video gaming platform, and in 2019 political candidates joined the platform, including Donald Trump and Senator Bernie Sanders. This particular live-streaming network does not allow political advertising, but the two candidates on the platform are connecting with at least some of the network's 140 million users.

And along with all the rapidly developing technology, there's still the basics. Let's start with that good old-fashioned snail mail. This is where direct mail comes into play. What's direct mail? Glad you asked. A pretty good chunk of Americans still rely on snail mail, and direct mail is unsolicited advertising that hits the mailbox. Okay, fine, it's a fancy way of saying junk mail. According to US Post Office research, 44 percent of millennials pick up mail at least six days a week, as do 60 percent of Gen Xers and 73 percent of baby boomers. Hence, this is a good space to catch voters and potential voters. But the materials should authentically resonate with the community. Mail that dubiously looks like an exclusive invite to the Essence Festival will likely get more attention than a stale campaign mailer featuring a picture of the candidate kissing babies on the trail. Interestingly, I have found that both Democratic and Republican candidates make appeals specific to Black folks only when they're very unlikely to be viewed by white voters. There's a bipartisan hesitance to publicly court Black voters, because that outreach may harm candidates' existing electoral coalitions with everyone else. So essentially Black voters are the dutiful wife who has been keeping candidates in power. But like pre-Kim/MAGA Kanye says, "But when you get on, you leave my ass for a white girl."

There's also that good old-fashioned face-to-face. Hit the streets and canvass. According to Garnett, the best tactic is still building interpersonal relationships. "Meet us at our doors. There are still lots of families in the Black community living in multigenerational homes—mama, grandmama, kids home from college, all living in one household." A Pew Research study found that a record 64 million people, or 20 percent of the US population, lived with multiple generations under one roof in 2016. "It's a multilayer sandwiched approach. From television, to radio, to satellite radio, cable, broadcast, to applications, to whatever foolishness is going on Black Twitter," Garnett continued. "You have to meet people where they are." But she says that when voters are asked in focus groups how they like to receive information, no one can say. They just know they want to have access to it. "So we know that means we have to get at them multiple ways. One piece of information isn't going to get it. They need to see things four, five, six times before they actually go through it and it actually resonates."

Aided by media misperceptions, national candidates often talk about Black voters as though we are all living below the poverty line in inner cities across America. This is demonstrably false—39 percent of African Americans live in the suburbs, 36 percent live in cities, and 15 percent live in small metropolitan areas, according to Brookings. And despite the notion perpetuated by the media that rural voters are almost exclusively white, this is also false. A whopping 10.3 million people, one-fifth of rural America, are people of color—meaning 10 percent of the Black population lives in rural communities. As I've stated, political interests are largely informed by geography. So reaching out to these voters must be tailored. In 2016, Black voters in rural places voted more like nonwhite voters nationally than they did like rural voters nationally.

Really, there's only one way to learn how to reach out and organize Black people, and this is important. Ready? Okay: LISTEN TO ACTUAL BLACK PEOPLE! This is why employing and empowering Black people in political spaces (and really every space) is integral to the success of campaigns and, as I've been reiterating throughout this book, democracy. And just to be clear, this is not to suggest that all Black people should be relegated to doing Black things. Our Blackness is just an *added* area of expertise, not our only area of expertise. Note that when Louisiana Governor Edwards realized he was in trouble with Black voters, he looked to Black staff. And when he didn't have the right people, he hired consultants. Black consultants—they matter. Billions are spent on campaign consultants—strategists, pollsters, direct mailers, and fundraisers who do the backstage work of modern campaigns. A 2019 survey of political firms conducted by *Campaigns & Elections*—a trade magazine covering political campaigns—found that 84 percent of respondents thought their companies were doing enough to promote diversity. They were not, as evidenced by the second finding in their research. Roughly half admitted that they had "witnessed or encountered racism" in the campaign industry, and a third of professionals agreed that perceptions of the political industry prevent more minorities from entering the field.

When I interviewed rapper Luther Campbell in Miami in 2019, he said, "Politicians aren't talking to the average Black person." He considers himself a bridge between politicians and the "average Black voter" in the battleground state. Regardless of how you feel about him, he has a point. Not enough people are talking to Black voters.

During the 2020 election cycle, there was a notable paradigm shift in the way candidates engaged Black voters. Democratic contenders introduced plans on issues like criminal justice reform, reparations, maternal mortality among Black women, HBCU funding, voter suppression, and systemic racism. But the media was still slow to catch up. During the primary, there were few times the candidates were asked about these issues on the debate stage. And when looking at the landscape and how elections are covered, I noticed there was a gap in the number of Black voters contacted by national media outlets. Certainly there are discussion panels on cable news outlets that, on occasion, discuss Black voters. Every so often they may include

Black voters in the discussion (shade). But segments highlighting studies and data focused specifically on Black voters—the largest minority voting bloc in the country—are few and far between. There are endless segments of campaign reporters visiting countless diners in the "heartland" where few, if any, Black people eat. Somehow these voters are supposed to represent what Americans are *really* thinking. I have never seen these reporters visit a wing stop in East Cleveland, Ohio, or Golden Corral in Montgomery, Alabama, after a Black Baptist church service, or a school parking lot in southwest Atlanta, Georgia. Are these not American voters in the heartland? Do their views not matter? Do they not represent part of the rising majority of this nation?

Moreover, white voters have the benefit of being discussed as multilayered communities, with their data constantly being disaggregated: college-educated white women, college-educated white men, suburban "soccer" moms (euphemism for white moms), Christian evangelicals (the chosen descriptor for white people who attend church), NASCAR dads, farmers, and so on. With over 41 million Black people in America, all of these descriptors could easily be applied to large Black populations. However, what we mostly hear in the echo chambers is "the Black vote," as if that captures who we are or what we care about. What about those former twenty-something men who used to play pickup basketball after work but are now dads dropping off and picking up kids at sporting events on the weekend? They can be stockbrokers, FedEx delivery guys, writers, dentists, middle managers, real estate agents, and so on. Let's call them basketball dads. What about the "get off work at five, I'm so sick of these white people at the office messing with me" Black mom? Or the single, childless Black women who are college-educated, high earners, well-traveled, professionals—"the aunties." Or the older Black men who have endured so much racism it wrinkles their faces, colors their hair, and informs their vote—"the Barbershop veterans." I've got dozens more. We don't all live under one umbrella. But no one has put the effort into coming up with cute colloquialisms for the Black community the way they have done for white people.

In October 2019, the *New York Times* ran an article describing an Elizabeth Warren campaign rally at an HBCU in South Carolina: "While Ms. Warren is rising in polls thanks to support among liberals, women, young people and college-educated whites, black voters, who are the most essential part of the traditional Democratic coalition, have yet to embrace her in large numbers," the reporter wrote. While the author acknowledges how critical Black support is to any candidate, the writing implies that at that time no Black voters supported Warren. Yet imagine if the data were broken down for Black voters as it was for every other group. Black voters too can be liberal and young and college-educated, and more. Perhaps college-educated Black women supported Warren, but working-class Black voters did not? Or maybe the inverse was true? We'll never know, because we are rarely afforded that lens.

So it's perfectly understandable that large swaths of people say that no one in the media, no politician or pollster has ever asked them what their lives are like. It's true. I've been voting for nearly twenty years, and I have never been polled. Democratic pollster Cornell Belcher dismisses the entire notion of questioning polls. He says, "When you go to the doctor, they don't take all your blood. They take a sample. Yet from that sample, they can still give an accurate reading of what's happening in your body." It's a fair point. But I know so many Black voters who have never been sampled, and too often our perspectives are not reflected. I should note here that every social scientist I spoke with for this book scoffed at the very idea of questioning public opinion polls and stressed that I needed to understand them better. So I set out to do exactly that. I discovered it's pretty easy to misread what polling represents.

Leading up to presidential elections, the media becomes obsessed with polls. It can give the impression that each pollster spoke with tens of thousands of people, and here are the results. Not so. "When you do a poll, typically a sample size is four hundred people or eight hundred, or one thousand, or if it's a really, really massive poll maybe two thousand people," said Fernand Amandi, the managing partner of Bendixen & Amandi, a multilingual and multiethnic public opinion research and strategic communications consulting firm. Also, sometimes the margins of error in these polls (the amount of random sampling error in a survey's results) is considerable. But that caveat is rarely mentioned in broadcast media. "Polling is a science just like engineering and medicine," he says. "There are certain variables and controls that you can apply to make sure that they are as representative and as comprehensive as possible. But also acknowledging and recognizing that there's always a margin of error as well. So the more robust and larger the sample size is, the more representatively designed in terms of what area you're looking at, or who you're targeting, the more confidence you can have in their results." And just because there are some bad polls, or badly designed polls, does not mean that all of them are poorly designed, Amandi says. It varies from poll to poll.

But where are pollsters finding these people? Personally, if my cell phone rings and it's a number I don't recognize, I'm not answering. "Is it a little bit harder to pull now, from the standpoint of cooperation rates? Yes," Amandi says. "You know, in the past, more people were willing to cooperate. Cell phones and caller ID and a lack of landlines, and there are some people who are completely offline—that makes it more challenging. But what we have found is that it does not impact opinions and perspectives." When phones became omnipresent, he says, pollsters were able to get direct access to people to participate in public opinion polling studies. Pollsters could make six thousand calls to get six hundred people to participate. "But today, you might need to make sixty thousand calls to get six hundred people to cooperate. So it just means it's more work for the pollster because of the cooperation rates."

But he says so much of this depends on who pollsters are going after—such as Black voters. Amandi conducted a study of Black voters in Miami in 2019 and explained how it worked. "We have a list of every single registered Black voter in Miami Dade County. They have a phone number assigned to their registration. So what you do is you literally just call through the list. And then what emerges from the number of interviews you do—if they're big enough, you will have a representative sample." But he cautions that it can get complicated when considering the demographic trends, and pollsters have to be careful not to get a false read. "Do you have enough women? Do you have enough men? Do they break apart in terms of country of origin? Like in Miami, we have a lot of Haitians. So if 40 percent of the people you're talking to in your sample in Miami are Haitian, then that's not a representative sample because Haitians only make up about 15 or 20 percent of the Black population in Miami Dade County voters."

And the best poll is only a snapshot of public opinion at that particular moment in time; it is not an eternal truth. So slapping a salacious headline on a poll like "Candidate X Takes Huge Dip in the Polls" should be viewed through the prism of when the polling took place. If you're not a person who follows politics, and you're polled during the week Trump's Ukrainian scandal blanketed every outlet with wall-to-wall coverage, by osmosis you may have heard Joe Biden's name mentioned again and again. That may influence your opinion one way or another simply by osmosis. Most network-employed data geeks (no shade, I'm into it) don't look like anyone in the rising majority, and that's often reflected in their interpretations of data. Every network obsessing over polls should have a Belcher or Amandi on staff. Polling minority communities is its own science. Most people don't understand statistics, and what twenty-four-hour press coverage fails to do is to explain how polling works. And by failing to cover us and poll us directly, the media excludes us from democracy. Very few polls have focused exclusively on Black voters on a massive scale.

That's exactly what Black Lives Matter founder Alicia Garza did. In May 2019, she released the Black Census Project, the largest survey of Black Americans since Reconstruction. They trained more than a hundred Black organizers and worked with some thirty grassroots organizations to reach more than thirty thousand Black people across the country, who are frequently sidelined in traditional surveys and mainstream politics. "We talked with black people who live in cities and in rural areas; black people who were born in the United States and who migrated here; black people who identify as lesbian, gay, and bisexual; black people who are transgender and gender-nonconforming; black people who are liberal and conservative; and black people who are currently and formerly incarcerated." Black women made up 60 percent of respondents due to the critical role they play in political and electoral organizing. Nearly 30 percent of those surveyed were between the ages of eighteen and twenty-nine.

Fifty-two percent of respondents said that politicians do not care about Black people, and one in three said they care only a little. Despite that, nearly 75 percent of respondents said they voted in the 2016 presidential election, and 40 percent reported helping register voters, giving people a ride to the polls, donating money to a candidate, or handing out campaign materials. Again, self-preservation trumps the patriotism that so many try to assign to civic engagement. "The Black Census shows that the Black electorate wants policies that improve our lives, not pandering photo ops at Black institutions," Garza said. It's interesting that so few of these people have been featured on cable news as voters to speak about what matters to them. Maybe they're not eating at the right diners in Iowa. "Campaigns that fail to understand or try to remedy the ways structural racism damages Black people's lives are doomed," Garza wrote in an op-ed published in the *New York Times*. "Without this analysis, their solutions will always miss the mark when it comes to Black voters."

Can we hold Black people accountable for not being engaged enough to seek out information on their own and exercise their civic engagement? I've obviously been adamant that campaigns must do better. And Jamal Simmons, a longtime political operative and on-air talent with The Hill TV, told me the blame is never with the voters—it's always the candidate. So it's an unpopular opinion I'm going to argue here, but—hell yes, we have to blame voters. At least for some of the challenges we have. People must play a role and take responsibility for our democracy. And too much of the American electorate has shirked that responsibility. Black people have more to lose, so it is in our own interest to become a part of this system so that, in the interest of the greater good, we can disrupt it. There are systems and people in place that depend on Black voters to abstain. We cannot be an unwitting coconspirator in these efforts.

Hashtags and Headlines

Given what's happening at every layer of US government and the media's failures in covering issues that impact Black people, we have an added responsibility to keep ourselves informed. We can all get on our high horse and proclaim Trump supporters to be uninformed, but I find this to be true on all sides of the divide. During the Obama administration I had many conversations where people both applauded and chided the president for successes or failures that were clearly not in his purview. Many people I know seemed to not know the difference between federal, state, and local government or who was responsible for what. But that never stopped anyone from publicly professing an opinion on social media. Nor did it seem to move anyone to want to understand. Not when politics pollinated pop culture.

Remember #ReclaimingMyTime? The phrase is actually pretty standard in congressional procedure. But Democratic California congresswoman Maxine Waters popularized the phrase when she used it in 2017 while questioning Treasury secretary Steve Mnuchin on why his office had not responded to a letter from her office regarding Donald Trump's financial ties to Russian banks. As she was questioning Mnuchin, he repeatedly interrupted her. Waters packed so much in the phrase, "reclaiming my time." When Black leaders, activists, and politicians speak, they do so from a place of love, from democracy denied. We have a way of taking phrases, actions, and activities that were previously flat and filling them with life and passion. Mnuchin was not just trying to deny Waters her authority as a sitting member of Congress; he aimed to dismiss her. She would not be denied. It was a reclamation of power in a public way against an administration whose very existence was based on denying Black people a voice and a chance to influence representative government.

"Reclaiming my time" became the hashtag used 'round the world. It began to pop up across all of my social media timelines in posts from followers and friends alike. "Monday is here already? Bring the weekend back. #ReclaimingMyTime." "Please don't ask me questions; you can google the answers yourself. #ReclaimingMyTime." And, "I'm about to put my out of office on, catch this flight, and see y'all in a week. #ReclaimingMy-

Time." Given all the symbolism of this phrase, the actual substance of it was also important. But many people didn't know
the actual story behind it. Large swaths of tweeters were reclaiming their time but had no understanding of the issue at
hand. Waters had requested that the Treasury Department provide to the House Financial Services Committee any records
that detailed Trump's financial ties to Russia, as well as those of
his family members and associates. Mnuchin never responded.
But none of that fits into 240 characters, and it doesn't come
with a clever hashtag so . . .

Sometimes we are the ones spreading disinformation. And
other times, we don't know what we're sharing at all. Some of
my Facebook friends would share news articles having only read
the headline, not the actual article. Sometimes their posts, accompanied with an endorsement, made the opposite point from
that expressed in the article. One friend shared this post days
after the 2016 Women's March: "Mike Pence Disappointed in
the 200,000 Husbands and Fathers Who Permitted Women to
Attend March," adding, "Yeah—that's #MAGA for you. As if
women can't decide for themselves." It was an article from *The
Onion*, which she didn't realize was satirical, a clear sign that she
had not bothered to read it.

But here's the deal: it's hard and lonely to be that one friend
constantly fact-checking. Must I always point out when friends
get things wrong? Must I always note the nuance my apolitical
friends miss when they declare an opinion on something political? Do I have to introduce a pointed history lesson about
the impact of colonization and how European goods, ideas, and
diseases shaped the changing continent of America, when they

were just trying to have a lighthearted conversation about how fashionable the ladies were in *Black Panther*'s Wakanda? Yes, I must. I'm that friend. But this, I discovered, is the easiest way to get muted in the group chat. So I've learned to pick my battles. What's a girl to do?

The incorrect assumption sometimes made in the media landscape is that people everywhere consume the minutiae of politics just as news producers do, and clearly they do not. News is easier to consume in terms of access—it's literally at our fingertips. But when it does not center us or our politics, it leaves us out of the democratic discourse. We must take the added step of staying informed and viewing and filtering our newsfeed through the prism that much of it is constructed by narrators either unfamiliar with or indifferent to our space in the American body politic. Meanwhile, there are droves of people across the divide who remain both willingly uninformed and passionately politically active. Fools with a cause can be just as dangerous as those without one. Fools are fools. When I have seen some interviews take place at Trump rallies with people declaring blind allegiance to the MAGA leader and espousing some of the most ridiculous and hateful things, I am baffled that their vote counts as much as mine. But, inspired by white supremacist lingo and policy, they are voting.

I grew up below the poverty line, so I know the 'hood. I live in a neighborhood of Washington, DC, that is surrounded by public housing and gentrifiers. As such, I will sometimes stop on the block and chat with the longtime neighborhood residents that gentrifiers pass by cautiously, never bothering to speak to their neighbors. When I talk to these folks who live below the poverty line, I am disappointed when they are so disgusted by the system that they choose to abstain altogether. But I understand their feelings of hopelessness when it comes to politics as it relates to their lives. We must give Black voters something to vote for. Obama was a once-in-a-lifetime candidate, and after his departure from the White House, some Black voters became dispirited with politics, feeling like Obama did not do enough to impact their lives.

Others became so used to being excited, they expected goose bumps from every candidate they encountered. As the country is still discovering, that's not going to happen. Every stump speech is not going to be comparable to a Sunday morning southern Baptist church sermon. And every candidate is not going to be a Martin Luther King Jr.–type orator. And policy discussions will not always be exciting and easy to comprehend. To inspire Black people by the masses, we must reimagine America. And in the minds of many people in the media, their vision of what America was, is, and should be is simply not ours.

14

Fade to Black

When I began the process of writing this book, I honestly wasn't sure where I'd land. I spoke confidently to my publisher, literary agent, and colleagues about how I was going to take a historical look at the role Black people have played in shaping democracy and apply that to how we could shape America's next phase. That's a good read, right? Well, what I discovered is—that's more like an encyclopedia. At best, it could be a series of books with volumes upon volumes looking at different time periods and subject matter. Black people's contribution to this country is so great that one century's worth of input is too voluminous for one book—but if someone attempts to write it, I will definitely read it.

So choosing instead to hone in on the two areas that have had the most impact on my life and, I contend, democracy at large—representation in media (or lack thereof) and how it influences our politics—was a cathartic process. Revisiting times in my professional past helped fuel my righteous anger today. Like most Americans, I was concerned when the country, with help from a foreign adversary and domestic voter suppression, turned the nation over to an unapologetic white supremacist. Like many, I was disappointed at how the media landscape responded. But I was never surprised. We are almost always expected to dilute our truths for the comfort of others. It is assumed that conversations will center white people and never communities of color. Predictably, the power brokers were out of step with the rising majority, and the assumption was that we would wait patiently until acknowledged. Not today.

The concept of white America being a minority has shaken the white male patriarchy (and its white female enablers) to its core. From lynchings to beatings to redlining and calling the cops at Starbucks, history has shown us that frightened people do desperate things. They don't care about playing by the rules. That can include silencing people whose truth they find incorrect, offensive, or controversial. But if we are to disrupt a space controlled by white media gatekeepers who are bleaching the Black American experience and destroying democracy at the same damn time, we must be bold enough to not be silenced. Our silent obedience will never buy us favor with the white male patriarchy that still dominates so much of what happens in the lives of people of color.

We, as Black people, are well positioned to be the great equalizers. I love being part of this secret society existing before their very eyes. We speak a language in which they recognize the words, presume to understand them. But the duality of our kind is hard to capture. There is something so ingrained in the souls of Black folk that it can't be taught or absorbed by osmosis. It's empowering. It has afforded us the audacity to change the lineage of the presidency, shape the legislative branch of government, draw the ire of the highest-ranking white supremacist in the land in the process, and—still unfazed—harness power so potent that it stretches the continent and dares these United States to take shape without our input. Our spirit is sewn into the fabric of America. There's room for equality and coexistence with all people. But not for people who aim to sideline our participation in this representative government.

So if it's a battle they want, let future generations say that we were the ones who stood on the frontlines, looked the opposition square in the face as we locked arms, and dared them to deny us our rightful place in the American body politic. We can set this democracy on course so that these United States can fulfill the promise of all that we hoped she could one day be. The candor in this book may leave some gatekeepers troubled. Consequence can sometimes accompany honesty. But the alternative is no longer palatable. We can let this country fall or bend this democracy to encompass the role Black people played in its formation. I choose the latter. The resistance is to be expected. But we all know what happens to things that don't bend. For generations we balled our fist and spoke our truth and "said it loud." And then the needle would move a bit. This go-round, it's time to say it louder.

We've got next.

Acknowledgments

Thank you to my mother, Ramona L. Cross, for raising me to be a reader and a writer.

Thank you to Albert Sanders Jr. Esq. for listening to me incessantly as I read you paragraphs and panicked about life on a daily basis. Dr. Jason Johnson, you literally saved this book. It would not be what it is without your sharp editorial input and the countless hours you spent picking it apart. Thank you to Dr. Richard Fordjour (Ricky) and Dr. Patrice Johnson (Trice) for always believing in me and for being constant voices of support.

Thank you to Joy-Ann Reid for believing in me so much and for talking to me about my book before a concept was ever in the works. Thank you Angela Rye for being such a loyal friend, saying to me "sis, I got you" and truly meaning it. Thank you to all the members of the Wakanda Council and, by extension, the Dora Milage (they know who they are): our modern-day Black Algonquin Roundtable is fertile ground for intellectual exchange and I am honored to share the space with all of you.

Thank you to Amistad, HarperCollins, and the entire team.

Thank you to my editor, Patrik Henry Bass, for journeying through this process with me. Thank you to my WME team Mel Berger and Henry Reisch for believing in me and helping to bring my goals to life.

Thank you to the Institute of Politics at the Harvard Kennedy School and the entire team. Thank you Lydia Carey for your timely research and fact checking. Thank you to my entire family—especially my uncles (Douglas, Brady, and Marco) for always making me feel like my voice deserved to be heard. To my brothers (Twan, Chucky, and Donnie), all my love as I pray you chart a meaningful and fulfilling course in life. I'm cheering you on. Thank you Cleve Mesidor, Robert Raben (thank you for being the only person to financially invest in *The Beat*), Brenda Arredondo, Liz Montano, Robert F. Smith, Dr. Michael Eric Dyson, Soledad O'Brien, Ta-Nehisi Coates, Terri Martin, Jade and Melissa Lopez, Scott and Carmen Craig, Gerald Walker, the Cougars (thank you for helping to grow this kitten), the GRITS ladies, and all the ladies of Overbooked.

Thank you to LaTosha Brown who is fighting to save democracy one race at a time.

And lastly, thank you to all the beautiful writers whose words and prose first sparked intellectual thought in my head. I am because of you.

About the Author

Tiffany D. Cross is a resident fellow at the Harvard Kennedy School's Institute of Politics. As an on-air political analyst, she is a longtime cable news veteran. Cross is a former associate producer for CNN, DC bureau chief for BET Networks, and liaison to the Obama administration. She cofounded *The Beat DC,* a daily newsletter intersecting politics, policy, business, media, and people of color that was widely read by Beltway insiders and media influencers. She attended Clark Atlanta University and lives in Washington, DC.